Mathematics Education Library

More information about this series at http://www.springer.com/series/6276

Edward Silver • Christine Keitel-Kreidt
Editors

Pursuing Excellence in Mathematics Education

Essays in Honor of Jeremy Kilpatrick

 Springer

Editors
Edward Silver
School of Education
University of Michigan
Ann Arbor, MI, USA

Christine Keitel-Kreidt
Psychologie
FU Berlin, FB Erziehungswissenschaft und
Berlin, Germany

ISSN 0924-4921   ISSN 2214-983X (electronic)
ISBN 978-3-319-11951-9   ISBN 978-3-319-11952-6 (eBook)
DOI 10.1007/978-3-319-11952-6
Springer Cham Heidelberg New York Dordrecht London

Library of Congress Control Number: 2014955213

Printed on acid-free paper

Springer is part of Springer Science+Business Media (www.springer.com)

Contents

**Part III   The Interaction of Theory, Practice and Politics
in Mathematics Education**

Introduction

Edward Silver and Christine Keitel-Kreidt

In 2007 Jeremy Kilpatrick was awarded the Felix Klein Medal honoring lifetime achievement in mathematics education from the International Commission on Mathematical Instruction (ICMI). Shortly after he received this prestigious award, we were approached by Alan Bishop, series editor of Springer's *Mathematics Education Library*, to inquire about our interest in editing a volume that would celebrate Jeremy's many professional accomplishments. We enthusiastically agreed, and shortly afterwards solicited the participation of a number of Jeremy Kilpatrick's former students and colleagues. Though it has taken longer than we had hoped to produce the volume, we are delighted that it is now available, and we hope that readers will benefit from the perspectives offered herein.

Ten of Jeremy's former students and colleagues from across the world have written thoughtful and informative chapters on a variety of topics all connected in some important way to Jeremy Kilpatrick. The volume reflects the myriad ways that Jeremy has influenced mathematics education and mathematics educators across his career—as a colleague, collaborator, friend, mentor, scholar, critic, and professional role model. His key role as global ambassador for American mathematics education and as conveyor of global perspectives to US researchers is evident in the fact that six of the ten authors hold professional appointments outside the USA, in Australia, Europe, Scandinavia, and the Middle East.

The chapters, however, are more than testimonials to Jeremy Kilpatrick. We all know that Jeremy would not be happy with a published volume that did not try to contribute in other ways to the advancement of mathematics education. Throughout his career he has been a tireless advocate for high-quality mathematics education research and practice. At times, he has been our field's harshest critic; at other times, its biggest booster. In that spirit, the authors who have written for this volume take up important issues in the field, treat them with great care, probe them critically, and seek to enhance professional dialogue on these issues.

The first section of the book contains chapters that focus on Jeremy as a colleague, mentor, and collaborator. Jim Wilson, who is Jeremy's long-time colleague from their days as graduate students at Stanford University to their current association as faculty colleagues at the University of Georgia, discusses highlights of their

50-year association as colleagues, collaborators, and friends. In additional to being a moving, personal tribute to Jeremy Kilpatrick, Wilson's chapter also provides a series of snapshots of the important developments in mathematics education over the past five decades or so in the USA—a history that Kilpatrick and Wilson have had key roles in creating and writing.

Patricio Herbst, a former student of Jeremy's at the University of Georgia and now a professor at the University of Michigan, devotes his chapter to an inquiry into and analysis of Jeremy's approach to doctoral dissertation advising and mentoring. Over the years, Kilpatrick has advised or served on the doctoral dissertation committees of hundreds of doctoral candidates in the USA (at Teachers College, Columbia University and at the University of Georgia) and abroad. Herbst describes Jeremy's mentoring approach, which is far less directive and advisor-centric than that of many others in the field, and probes Kilpatrick's rationale for adopting this approach. Herbst contrasts Jeremy's approach with his own views on the dissertation advising process. In so doing, Herbst uncovers the nuanced ways in which each of these apparently different approaches to doctoral mentoring can succeed in supporting the development of a new generation of scholars capable of conducting rigorous research on important issues.

In her essay, Christine Keitel—a professor at the Freie Universität in Berlin and a frequent collaborator with Jeremy on projects, publications, and other professional activities—relates an anecdote about how her first encounter with Jeremy Kilpatrick at a conference at the University of Bielefeld led to their eventual collaboration with Geoffrey Howson on a seminal book on mathematics curriculum development that was published in 1981. This story reveals not only the keen intellectual curiosity that those who know Jeremy always feel when they interact with him but also the supportive colleagueship that those who are fortunate enough to have worked with him have felt in those interactions.

The next section of the book contains chapters that address aspects of two major themes that have been prominent in Jeremy Kilpatrick's scholarship over the years: mathematical problem solving and mathematics curriculum. In his contribution, John Mason, emeritus professor at the Open University (England), contrasts his perspective on mathematical problem solving with that of Jeremy's, while probing a time-honored question of interest to the field: What constitutes a mathematics problem? Though this question was vigorously debated during the 1970s and 1980s, Mason offers a fresh perspective, contrasting problems as they appear in curriculum materials and problems as experienced by solvers.

João Pedro da Ponte, a former student of Jeremy's and now a professor at the University of Lisbon, offers a historical perspective on mathematical problem solving. Inspired in part by Jeremy's writings on the history of research in mathematics education, he traces the appearance and use of mathematical problems in school textbooks in Portugal. Drawing on other work by Kilpatrick, and also Kilpatrick's mentor George Pólya, João Pedro da Ponte provides a framework for distinguishing among different types of tasks contained in textbooks: exercises, problems, investigations, and explorations. His analysis of three generations of algebra textbooks reveals several interesting trends in the evolution of the nature of the tasks provided to students in the books.

In his chapter, Thomas Lingefjärd, another of Jeremy's former students and now a professor at the University of Gothenburg in Sweden, offers a very different perspective on problem solving. Picking up on the theme of problem formulation that Jeremy has written about in several of his papers, Lingefjärd examines geocaching, which is an outdoor recreational activity, in which participants use a global positioning system (GPS) receiver or mobile device and other navigational techniques to hide and seek containers, called "geocaches" or "caches." He illustrates how the combination of open-ended investigation and technology in geocaching capture many of the key features of problem solving that have been identified in the literature. In so doing, he illustrates how the tradition of the amateur mathematician, long evident in history of number theory, might live on in this contemporary recreational activity.

In her contribution, Vilma Mesa uses ideas and perspectives from Kilpatrick's writings as the basis for her analysis of the mathematics curriculum found in the modern American community college. Mesa, who met Jeremy in her native country of Colombia, became a student of his at the University of Georgia, and is now a professor at the University of Michigan, contrasts the notions of curriculum as written in documents and curriculum as enacted by teachers and students to identify some of the tensions embedded in the historical, societal, cultural, and political conditions that shape contemporary community colleges in the USA. She shows how these tensions support the status quo and impede well-intentioned efforts to reform the curriculum to improve learning opportunities and outcomes for students.

In the third section of the book, authors take up some other themes that have been treated by Jeremy Kilpatrick in his scholarly writings, particularly his advocacy for the importance of establishing a solid theoretical foundation for research conducted in the domain of mathematics education, his critical attention to the ways in which politicians and policy makers often use access to and the attainment of proficiency in school mathematics as a tool to maintain political and social conditions of inequity, and his critique of mathematics education as a professional field. These themes are all evident in Christine Keitel's chapter in this section. Taking a critical perspective, she traces the historical development of the view of mathematics as a highly specialized technical skill, a scientific discipline, and a political tool to exercise power and authority from the ancient Greeks to the modern interest in applied mathematics and in "mathematics for all."

In his chapter, Alan Bishop, a long-time professional colleague of Jeremy's and a professor emeritus at Monash University (Australia), also takes up the political dimension of mathematics in contemporary society. He offers an updated perspective on the theme of "mathematics for all" by focusing on tensions between the notion of democratizing access to mathematics and the widely accepted construct of mathematical ability. The bulk of his chapter takes a critical perspective on research on mathematics teaching and learning, with an eye toward identifying promising targets for research attention that could advance our understanding of the possibilities and challenges of democratizing access to mathematics.

In her contribution, Pearla Nesher, a long-time associate of Jeremy in international professional organizations and professor at the University of Haifa (Israel), takes up the contemporary use of theory in mathematics education research. She recalls a lecture given by Jeremy Kilpatrick at a PME conference more than 25 years ago on the strengths and limitations of constructivism as a theory to guide mathematics education research, and she revisits that issue as well considering critically the recent proliferation of theoretical perspectives that are now used by researchers in the field. She argues for the value of our becoming more insistent on using research evidence to test our theories rigorously.

In the final chapter, Ed Silver—who was a former student of Jeremy's at Teachers College, Columbia University—uses Shulman's "commonplaces of a profession" as a lens through which to view the career of Jeremy Kilpatrick. Although Jeremy has written that mathematics education is not a profession, Silver argues that Jeremy's scholarly pursuits and his range of activities in mathematics education illustrate important features of professionalism—including the obligation of *service* to others, the need for *understanding* of a scholarly or theoretical kind, the need for *learning from experience* as theory and practice interact, and a professional *community* to monitor quality and aggregate knowledge—that mark a pathway that mathematics educators could follow toward making our field a profession.

Over a career that has spanned 40 years, Jeremy Kilpatrick has made enormous contributions to mathematics education, not only in the USA, where he has been a leading figure almost from the beginning, but also across the globe through his involvement in many international endeavors. When a history of mathematics education is written in 25 years or so, there is little doubt that Jeremy Kilpatrick will be a central figure because of his important contributions to critical developments in the field, his steadfast advocacy for research of the highest quality, and his remarkable skill in balancing advocacy and criticism in all that he has done. He will surely be noted as a leader in the persistent pursuit of excellence in research and practice in mathematics education. One might wonder, however, who will write that history; who will fill the important role of historian in the field that Jeremy has played for so long?

Whether one thinks of Jeremy Kilpatrick as editor of the *Journal for Research in Mathematics Education* for six critical years, co-editor (with Izaak Wirzup) of the *Soviet Studies in the Psychology of Learning and Teaching Mathematics*, chair of the National Research Council committee that produced *Adding It Up: Helping Children Learn Mathematics*, co-author (with George Pólya) of *The Stanford Mathematics Problem Book: With Hints and Solutions,* winner of prestigious awards (e.g., Felix Klein medal from ICMI, Lifetime Achievement Award from NCTM), or in his many other roles and activities, it is clear that Jeremy's legacy is enormous, and his influence is still growing. Through his actions and words, Jeremy Kilpatrick has given much to the field of mathematics education. We are pleased to offer this volume to him as one small token of admiration and gratitude from the field for all he contributed.

Part I
Jeremy Kilpatrick: Colleague, Mentor and Collaborator

Chapter 1
Fifty Years and Counting: Working with Jeremy Kilpatrick

James W. Wilson

Abstract This is a personal reflection about Jeremy Kilpatrick by a friend and colleague. I attempt to portray him as an outstanding scholar in mathematics education but from a personal perspective as well as professional. We worked together as graduate students and research associates at Stanford, and our collaboration continued as he joined the faculty at Teachers College, Columbia, and I joined the faculty at the University of Georgia. In 1975, he joined our faculty in Georgia, and I was his department head for the first 18 years he was here. Jeremy's mentors in graduate student days included George Polya, Edward G. Begle, William Brownell, and Lee J. Cronbach. I submit that Jeremy's link to these four scholars is a unique and rich intellectual heritage, but he also earned the admiration and support of each of them. More than any other person, Jeremy Kilpatrick has transformed the field of mathematics education and led us into the development of our field as an emerging discipline. His vision and influence are recognized worldwide, and he has brought an international perspective to mathematics education.

Keywords Colleague • Working with Jeremy Kilpatrick

Introduction

Others will cover different aspects of the remarkable career of Jeremy Kilpatrick. I will drop all pretense of detached objectivity; we have been colleagues for almost 50 years. He is a coworker and friend, and my career has certainly been supported by his. Rather than provide a chronology of events, however, I will take you on a somewhat random journey. I want to provide some reflections that might allow readers to see a different, more personal perspective.

J.W. Wilson (✉)
Department of Mathematics & Science Education, College of Education,
University of Georgia, 290 Hampton Court, Athens, GA 30605-1404, USA
e-mail: jwilson@uga.edu

© Springer International Publishing Switzerland 2015
E. Silver, C. Keitel-Kreidt (eds.), *Pursuing Excellence in Mathematics Education*, Mathematics Education Library, DOI 10.1007/978-3-319-11952-6_1

3

I came to value the work of Jeremy Kilpatrick in many ways over the years. First, he has been and continues to be a mentor as well as a colleague. Second, he personifies excellence and scholarship. Third, he is a great colleague, always carrying more than his share of the load. Fourth, he inspires students and colleagues to have high expectations of themselves. Fifth, he is a friend and neighbor, always supportive. We have all been especially blessed that he has been at the University of Georgia to work with us.

The Person

There is the person who is a personal friend. Never in 50 years has he had a harsh word for me although I am sure I provided plenty of reasons for them. We share a love and appreciation for mathematical problem solving. Our casual conversations, when they get beyond family, cover many other interests.

There is the person who is a day-to-day colleague in a university department. He is always a team member who takes on the whole range of responsibilities and is involved in all aspects of our departmental mission – undergraduate, graduate, research, and service. He is a faculty member who treats his colleagues with kindness and respect.

There is the person with an even demeanor who is always a gentleman and a scholar. Professional discourse must engage in the exchange and criticism of ideas. Unfortunately, many in our profession today engage in a confrontational type of criticism that does not separate ideas from personal integrity. That is not the case with Jeremy. His criticisms can be incisive and sharp, but they are always directed toward the ideas and not the person. There is no element of hostile tone in his discourse. This demeanor enables him to be effective in any group, whether it is a faculty meeting, a committee meeting, or an elite national or international commission.

There is the person who is an outstanding mentor. A large cadre of former doctoral students attests to his skill as a mentor. He has also been a mentor of many other mathematics educators throughout our profession. Those who have been collaborators on various projects also recognize this mentoring talent.

This photograph shows Jeremy Kilpatrick in his Stanford University academic robes but the hat from his Honorary Doctorate, Gothenburg University. He holds an A.B. and an M.A. from the University of California at Berkeley, an M.S. and a Ph.D. from Stanford University, and an honorary doctorate from the University of Gothenburg in Sweden. Jeremy has taught courses in mathematics education at several European and Latin American universities and received Fulbright awards for work in New Zealand, Spain, Colombia, and Sweden. He was a charter member of the U.S. Mathematical Sciences Education Board and served two terms as Vice President of the ICMI.

Honors and Recognition

Member, National Academy of Education
NCTM Lifetime Achievement Award, 2003
ICME Felix Klein Award, 2007
Honorary Doctorate, University of Gothenburg
Associate, National Academy of Sciences
Fellow, American Educational Research Association
Regents Professor, University of Georgia
Over 50 doctoral students for whom Jeremy was the major Professor

Doctoral Students

One's doctoral students help define one's legacy in the field. So it is with Jeremy. He has been the major professor of over 50 doctoral students at Teachers College, Columbia, and the University of Georgia. Moreover, he has served on the advisory committees of many others. When he is on the advisory committee, he is never an idle observer. Instead, he has the unique talent to challenge the candidates to do their best, and his expertise augments the rest of the committee. Rather than list all of his doctoral students here, the record can be found on the web at the Mathematics Genealogy Project (http://genealogy.math.ndsu.nodak.edu).

At the University of Georgia, Jeremy's skill and even demeanor in being a mentor is well recognized. His EMAT 9630 Critique of Mathematics Education literature is a right of passage for almost all doctoral students. It is a time when students are given significant guidance from being consumers of knowledge to becoming scholars, the move from being consumers of knowledge to producers. No, the course is NOT required – but no one would want to avoid it. It is taught every fall term. When Jeremy was on leave one fall, I taught the course. When he returned the next fall, all but one of the doctoral students I had taught retook the course from Jeremy.

A Key Event

In the spring of 1969, a group of us gathered for a working session. We chose New York City. Those present were E. G. Begle from Stanford, Jeremy Kilpatrick from Columbia, Ray Carry from Texas, and Tom Romberg from Wisconsin, and me from Georgia. This is the team who had spent a lot of time working together at Stanford before each of us went off to assume faculty positions at the respective sites. Jeremy had been at Columbia 2 years, since 1967, and I had been at Georgia for 1 year. The meeting was for the purpose of working on reports for NLSMA (the National Longitudinal Study of Mathematical Abilities). There were other meetings of this group in those years, but this one became a bit different. There were many side discussions, but I remember Jeremy telling me I should accept the appointment as Department Head I had just been offered at Georgia. We had no thought at the time that Jeremy would eventually join the faculty at the University of Georgia.

Fast-forward a few more years. In 1973, Jeremy went on leave as a visiting lecturer at Cambridge, Larry Hatfield from Georgia was a visiting faculty member at Columbia to fill in for Jeremy, Robert Reys from Missouri came to Georgia as a visitor to help us. When Jeremy returned from England in 1974, he had a full beard and came to Colorado to work with Reys and Wilson on analyses of the National Assessment data. Meanwhile, Jeremy had been courted by Ohio University for the R. L. Morton Endowed Chair but declined. Len Pikaart of Georgia accepted the Morton Chair and that created a senior level faculty position at Georgia. In 1974–1975, Alan Osborne came from Ohio State to be a visiting instructor at Georgia, and I took a leave of absence to work at the National Science Foundation. We began a national search for a senior faculty member, and in the process, Jeremy Kilpatrick came to Georgia in 1975. In fall 1975, I returned from NSF and resumed my role as Department Head.

Jeremy's recommendation in 1969 that I become the department head gave me argument for saying "you got me into this, so come and work with us." For the next 18 years, I was his department head, and his many accomplishments helped make it look like I was doing something right. Eventually, HE was to become the department head – although briefly. Jeremy was named the interim Department Head in 2005 and served for 9 days. For seven of those days, he was out of town.

His other department heads at the University of Georgia for at least two 3-year terms have included Larry Hatfield, Patricia Wilson, and Denise Spangler.

Family

Jeremy and Cardee Kilpatrick were newlyweds when my wife Corene and I met them in 1962. Over the years, the families have remained close. There were many years while our children were growing – the two Kilpatrick boys and the three Wilson girls – that we shared Thanksgiving celebrations, either in New Jersey or in Georgia. As children went off to college, moved on to families of their own, those events have been replaced. The Kilpatrick home in Athens has always been a venue for students and friends to come to parties or visit. The renovations of their home on Woodlawn were epic. The families continue to share and celebrate either other's significant milestones.

Cardee was also a Stanford student. After moving to Athens, she became very involved in public service, serving on the Board of Education and then on the Board of Commissioners for two elected terms. Now retired from the Commission, she remains very active in Athens community activities.

Sons Judson and Bart attended Carleton College and Swarthmore College, respectively, and the picture here shows the family at Bart's graduation.

The Leland Stanford Junior University

Official title of our university! Jeremy and I were doctoral students at Stanford; he began his graduate assistantship there in 1961 so he was the experienced one when I arrived in 1962. Our first collaboration was to share the supervision on one high school intern for the Stanford Secondary Education Project. The SEP was an innovative (for that time) project bringing highly selected graduate students to Stanford to prepare them to teach in Secondary School. His assistantship was to supervise all of the mathematics interns. I was assigned to be a computer programmer on another project, and the director allowed me to help supervise one intern. I learned a lot about high expectations and intellectual support in that first collaboration. Between 1962 and 1963, we both finished coursework and passed the written and oral examinations. His orals were on February 4, 1964; mine were on February 6.

Stanford in the early 1960s was an exciting environment. Jeremy completed a master's degree in mathematics in 1962. Excellent faculty-taught courses in our doctoral program and the on-campus interaction with other full-time doctoral students were a great learning experience.

In those 2 years, we both participated in the weekly mathematics education seminars run by Professor Begle and in the Woodside Seminars run by Professor Coladarci. These seminars, in retrospect, were the heart of our experience as doctoral students, and Jeremy stood out as a fellow student who had keen insight, leadership, intellectual merit, and the ability to help all of us learn. The Begle seminar involved all of us in mathematics education; the Woodside seminars were by invitation and specifically constructed to get students from many areas of education involved.

We came to recognize, however, that the real essence of doctoral study was the apprenticeship opportunities we had in working on projects or assisting faculty. For Jeremy, the opportunity to be Professor Polya's graduate assistant with the Freshman Mathematics Seminar was unique and one that he has always valued.

Over the years, our concept of apprenticeship as a component of doctoral study has evolved, and although we could not verbalize it with that term, our work with SMSG and the NLSMA project provided that dimension of doctoral study.

Before going on to that side of our doctoral experience, however, it is worth a comment about dissertation projects. Many outsiders thought that doctoral student staff at SMSG would just take a bit of the NLSMA analyses (or some other project) and make it into a dissertation. Rather, we expected, and Professor Begle expected, our dissertation development and research to be independent of the projects. This is particularly significant for Jeremy because it enabled him to develop a dissertation study that was to greatly impact future research in the field of mathematics education.

SMSG Headquarters and NLSMA

We each became full-time Stanford employees at some point, to work for the Headquarters Staff of the School Mathematics Study Group—along with Ray Carry, Tom Romberg, Jerry Becker, Gordon McLeod, Fred Weaver, Mervyn Dunkley, and others on the Research and Test Development (later called Research and Analysis) team that I headed. All of them except Gordon and I went off to university positions in 1966 or 1967, but we all continued working on the reports of the National Longitudinal Study of Mathematical Abilities, reconvening in the summers through 1972. Jeremy was instrumental in enabling this team, and others who joined us, to accomplish the tasks we set out to do. Some 40 volumes of reports were assembled.

Jeremy actually began working with NLSMA, with an office at The Elms rather than Cedar Hall, in 1962. Len Cahen was the project manager with NLSMA. NLSMA was funded by NSF to follow approximately 115,000 students, beginning in Grade

4, Grade 7, and Grade 10, for 5 years. Initially, it was thought that appropriate batteries of tests could be used for testing fall and spring of each year by using existing tests for easily creating some new ones. Advisors and staff quickly realized the project would need to have extensive instrument development activities.

The Begle seminar devoted a lot of time to conceptualizing appropriate measuring of mathematics achievement. All of us struggled to understand Bloom's taxonomy, or adapting it, to map out appropriate measures of mathematics achievement. The National Assessment Project eventually adapted versions of our "model."

Jeremy and Len were at the core of helping us conceptualize the issues and finding the expertise to deal with instrument development as well as handle the logistics of distributing, administering, and scoring instruments twice a year for 115,000 students. We relied heavily on the advisory board for NLSMA—called the panel on tests—and consultants. Members of the panel were:

Edward G. Begle, director of SMST, Stanford University
Richard Alpert, Harvard University
Max Beberman, University of Illinois
Robert Dilworth, California Institute of Technology
Jerome Kagan, Harvard University
M. Albert Linton, Jr., William Penn Charter School
William Lister, SUNY Stony Brook
Samuel Messick, Educational Testing Service
J. Fred Weaver, University of Wisconsin

Consultants were:

Arthur P. Coladarci, Stanford University
Lee J. Cronbach, Stanford University
R. Darrell Bock, University of Chicago
David E. Wiley, University of Chicago
John C. Wright, University of Minnesota

The panel members and the consultants were active participants in the research effort. The headquarters staff had a diverse set of experts to draw upon and often traveled to individual's home offices to work on project activities, whether instrument development, data management, or analyses. Often we had meetings of people with diverse background (e.g., mathematics, psychology, research methods) with very little turf for engaging in conversation. Jeremy emerged as an "interpreter" of sorts, often becoming the person to facilitate communication from one group to another.

The value of NLSMA to Jeremy, and to the rest of us, may have been much more in the opportunity to learn from this marvelous group of experts than in the reports we produced.

Subsequently, 40 volumes of NLSMA Reports were published. Jeremy was the author of several but integrally involved in the development of all of them. NLSMA had a vast amount of data on achievement and ability measures. Many more analyses were outlined and planned by the staff, the panel, and the consultants. NSF, however, was not interested in the funding of the proposed analyses, and, as time and careers moved forward, we all found other demands on our time. When I became a staff member at NSF, I encountered a hostile environment to research and evaluation in mathematics education and became aware of how unusual it was for NSF to fund NLSMA from the MIDS group (Materials and Instructional Development). Senior staff at NSF went out of their way to tell me they had opposed to the NLSMA funding. Clearly, the esteem NSF staff held for Professor Begle and SMSG overcame barriers to funding the implementation of the NLSMA project.

Mentors

Jeremy's remarkable career includes a combination of mentors that should be noted. First of all, his doctoral advisor was *Edward G. Begle*. Clearly, Jeremy was one of Professor Begle's most outstanding doctoral students. Ed Begle came to Stanford with a joint appointment in mathematics and education, but primarily as director of the School Mathematics Study Group, which transferred from Yale to Stanford in 1961. The immersion in the SMSG environment, with the many advisory committees and visitors, was very formative for Jeremy. Contacts from mathematicians, educational researchers, statisticians, and mathematics educators were the seeds for building relationships throughout his career. Professor Begle had great confidence in Jeremy's insight, scholarship, and knowledge about the many issues being confronted in curriculum development and research. The Begle seminar was the only class offered by Professor Begle, but it was truly a community of scholars, and the centrality of Jeremy's role in the seminars cannot be underestimated. Professor Begle certainly fostered and encouraged Jeremy's development of scholarship in those days, but it might be best described as "learning together." As a mathematician turned mathematics educator, Professor Begle was covering some of the intellectual terrain of mathematics education at the same time as his student.

In his Master of Arts degree in education at the University of California, Jeremy was a student of *William Brownell*—one of the most prolific researchers in the study of meaning with mathematics learning from 1930 to 1960. Later Jeremy worked on writing assignments with Fred Weaver, a mathematics education professor from Boston University and then the University of Wisconsin, who had been a student of Brownell's at Duke in the 1940s. Together, Weaver and Kilpatrick edited two volumes of William Brownell's theoretical and research papers.

Perhaps a most noteworthy mentor for Jeremy was *George Polya*. In the mid-1960s, Professor Polya reached age 75, and his only teaching was the Freshman Seminar in Mathematics at Stanford. Jeremy was his graduate assistant. Professor Polya also served on Jeremy's doctoral advisory committee, and the two of them published a book of the Stanford mathematics problems and the solutions. That book was out of print for a few years but subsequently republished in the Dover Classics series.

A third member of Jeremy's doctoral committee at Stanford was the psychologist *Lee. J. Cronbach*. Professor Cronbach was an advisory board member to NLSMA so Jeremy was getting his guidance both in the dissertation research and in the project. The demanding guidance Jeremy received from Cronbach in dealing with nonstandard methodology issues was tremendous. It showed Cronbach's capacity to think outside of his standard operating milieu and Jeremy's capacity to assimilating that guidance into some very new methodology for the field of mathematics education. I will say more about this later.

Begle, Brownell, Polya, and Cronbach all held Jeremy in very high regard. Given his talent, and the guidance of these mentors, Jeremy's outstanding accomplishments in his career were expected. For one individual to be impacted significantly by four such diverse giants in their fields is fortuitous, and all were pleased with Jeremy's accomplishments.

Teachers College, Columbia University

Jeremy became a faculty member at Columbia University in 1967 and quickly built an international reputation as a young scholar. At Columbia, he had the benefit of working with distinguished scholars from throughout education. In particular, he had the opportunity to work with Bruce Vogeli, Philip Smith, Paul Rosenbloom, Myron Rosskopf, Howard Fehr, and Robert Thorndike, and many others. He was particularly adept at guiding doctoral students, and he was an outstanding teacher. Between 1967 and 1975, Jeremy advanced in rank at Columbia and in the esteem he held in the field of mathematics education. At least 18 doctoral students completed their degrees with Jeremy as major professor between 1970 and 1975. He served on the advisory committees for many more.

University of Georgia

In 1975, the Department of Mathematics Education at the University of Georgia welcomed Jeremy Kilpatrick to the faculty as a senior (but not old) faculty member. He joined a faculty that included Leslie Steffe, Larry Hatfield, Thomas Cooney, Edward Davis, Joseph Hooten, James Wilson, William McKillip, Michael Mahaffey, Mary Ann Byrne, and Edith Robinson. Sigrid Wagner joined us in 1976 and Mary Anne Byrne left. This team remained intact for the next 10 years as collectively we built the mathematics education program to a level of national and international visibility.

Jeremy embraced every aspect of our departmental mission. This included significant work in teacher education. His teaching record, whether undergraduate, graduate, or combined, was excellent. This continues now 37 years into his journey with us. He has been the major professor for at least 34 former doctoral students and on the advisory for many more. Of course, he has brought significant recognition to the university for his many national and international activities and responsibilities. With all of his accomplishments, however, he has always been totally involved in the department and all of our programs.

Regents' Professorship

Bouncing off to another time, Jeremy was named Regents' Professor, University System of Georgia, in 1992. Putting together the recommendation and documentation for this appointment was one of my most enjoyable tasks in over 23 years as Department Head. It was also remarkably easy because Jeremy had created an outstanding career record. Outside evaluators included Alan Bishop, Geoffrey Howson, Thomas Romberg, Lynn Steen, Robert Orrill, Mogens Niss, Nicholas Balacheff, J. M. Fortuny, and Anna Sierpenska. Their message to the committee deciding on this appointment was one of universal acclaim.

The Regents' Professor appointment is bestowed by the Board of Regents on truly distinguished faculty at the University of Georgia whose scholarship or creative activity is recognized both nationally and internationally as innovative and pace setting. Only one award is made each year. It includes a permanent salary boost and a yearly academic support account. The initial appointment is for 3 years and is considered for renewal for a second 3-year term. After 6 years, the award becomes permanent.

Transforming Mathematics Education

It sounds almost presumptuous to argue that any one scholar has transformed the discipline. By his outstanding record of scholarship and his national and international leadership, Jeremy has done that. I want to argue here, however, that some specific personal accomplishments have, in fact, had considerable more impact, enough impact to alter significant aspects of our discipline.

Analysis of Verbal Protocols

In time from his doctoral work in the mid-1960s through the early 1970s, he had a major influence in redirecting methods and measures of scholarly inquiry in mathematics education. His work with the use of verbal protocols to develop traces of student processes in problem solving really transformed the field. This process was refined in his dissertation research and later in the dissertation research of his students. It opened up the widespread acceptance of such methods in mathematics education research.

Editorship of JRME

He assumed leadership roles in NCTM and from 1982 to 1988, he was editor of the *Journal for Research in Mathematics Education*, Volumes 13–18. JRME became an official journal of NCTM during his editorship and was firmly established as a leading outlet for research in mathematics education. He was the fourth editor of JRME after it was started by NCTM in 1970. His editorship was a time of transition for the journal to its recognition as the most important outlet for research in mathematics education, and it remains so today. In my view, Jeremy Kilpatrick was the unique individual to bring JRME out of its growing pains and make it into a top-ranked professional publication.

Bridging the Disciplines

Mathematics is an old and established discipline. Mathematics education is a new and emerging discipline. Jeremy Kilpatrick has participated in the leadership of mathematics organizations. For example, he was elected to the Board of Governors of the Mathematical Association of America and been at the table to help colleagues from mathematics develop understanding of the discipline of mathematics education. Many conversations have started with a mathematician saying to him "I just do not understand what mathematics educators do," and Jeremy has the respect in both disciplines to inform. He was charter member of Mathematical Sciences Education Board, served in many roles for the International Mathematics Union and its International Commission on Mathematics Instruction. He was the first non-mathematician to present a Plenary Address to the ICMI. His efforts have encouraged many gifted mathematicians to embrace concerns about mathematics education.

International Ambassador for Mathematics Education

The Felix Klein award from ICMI for outstanding lifetime achievement in mathematics education is perhaps the most prestigious recognition of international accomplishment and impact. Jeremy received that award in 2007. The award committee noted that he had worked in at least 11 different countries and was extraordinarily knowledgeable about the international literature in mathematics education. In dramatic fashion, he has been instrumental in bringing global perspectives, not just to the USA, but also to a worldwide audience. Early in his career, he was editor of a series of translations for Soviet Studies in Mathematics Education, bringing a vast and important literature to worldwide attention. He was also co-editor of the three ICMI International Handbooks on Mathematics Education.

Assessment

When the NCTM developed standards for mathematics assessment in 1995, it was quite appropriate that Jeremy Kilpatrick be in charge of the team writing the standards. He has published, advised, and spoken about mathematics assessment in various roles throughout his career. He has been involved in almost every major mathematics assessment project (such as The National Assessment of Education Progress). He has served on advisory committees for MAA, College Board, National Academy of Education, Educational Testing Service, National Research Council Board on Testing and Assessment, and the Organisation for Economic Cooperation and Development. He has influenced this area of interest in mathematics education by his knowledge of mathematics, awareness of mathematics curriculum and standards, publications of theoretical perspective, writing about and being involved in practical matters of instrument development, and providing valid interpretation of assessment results.

History of Mathematics Education

Many of his publications and presentations remind us of where we have been in mathematics education, not so much as a chronology of events, but rather the evolution of ideas and policies. He is attuned to theory and practice and the roots of these ideas. As a student of Brownell, Polya, and Begle, Jeremy had direct contact to some of the most influential ideas in the development of mathematics education, and he has been instrumental in moving us forward but at the same time keeping us aware of where we have been.

Moving Us Forward

Today, Jeremy Kilpatrick remains one of the most active and involved mathematics educators in the world. My head spins trying to comprehend all of the projects, advisory committees, teaching activities, and writing that he has in progress. One might expect some scaling back after age 50, but his case is one of increasing opportunity to lead us in the global development of the discipline of mathematics education.

With all of his travel and commitments, one might expect he is an occasional visitor here in the department. To the contrary, he remains a very active faculty member, teaching undergraduate and graduate courses, guiding doctoral students, serving on mentoring committees for younger faculty, and being involved in university governance. The Regents Professorship was awarded about 15 years ago, and he has made it a platform for expanded intellectual leadership. It is one of four excepted or named professorships we have in the department. At the core of his approach to being a faculty member, students are his priority. He is always accessible to any student, not just his own advisees, not "only by appointment," and students learn to seek and value his advice.

I am indeed fortunate to have had Jeremy Kilpatrick as a friend and colleague for these past 50+ years. The title of this piece says "working with" but the friendship is even more significant.

Epilogue

There is no epilogue—we are not done yet.

Chapter 2
What's Involved in the Work of Dissertation Advising? An Interview with Jeremy Kilpatrick and Some Personal Reflections

Patricio Herbst

Abstract This chapter addresses the mathematics education doctorate and, in particular, the dissertation or thesis. The core of this chapter is an interview with Jeremy Kilpatrick in an effort to document Jeremy's own advising style. The interview is situated within an attempt to bring out more general issues that are in play in how a mathematics education scholar might advise or direct a doctoral dissertation, fueled mostly by my own introspection into this role. After brief descriptions of dissertation experiences in other fields and personal descriptions of my own experience working on my dissertation, I document Jeremy's responses to the question of how he sees his style as dissertation advisor. I then propose that the dissertation work develops in response to four stakeholders: (1) the student, (2) the advisor, (3) the field, and (4) the institution. In an exercise of speculation, I examine those stakeholders in terms of how they could conceivably be invested in the successful completion of a doctoral dissertation. I go back to the interview with Jeremy Kilpatrick to see how Jeremy responds to the question of how and how much each of those stakeholders matter in his own dissertation advising style and how he views the dissertation.

Keywords Doctoral advising • Dissertation • Research culture

Introduction

The invitation to contribute to a volume in honor of Jeremy Kilpatrick was very appealing to me, but it prompted me to think what I could write about that would connect well with Jeremy's scholarship. While none of the research topics of assessment, curriculum, or problem solving that Jeremy has worked on connects very well with my own work, his mentoring does and in a very important way. I did my

P. Herbst (✉)
School of Education, University of Michigan, Ann Arbor, MI, USA
e-mail: pgherbst@umich.edu

© Springer International Publishing Switzerland 2015
E. Silver, C. Keitel-Kreidt (eds.), *Pursuing Excellence in Mathematics Education*, Mathematics Education Library, DOI 10.1007/978-3-319-11952-6_2

doctorate under Jeremy's direction at the University of Georgia. He has influenced me the most as a dissertation advisor. Explicitly, Jeremy taught me to write for the field. Tacitly, he has given me an example and a point of departure from which to start my own work as a doctoral advisor. My proposal to the editors was to interview Jeremy about his work as a dissertation advisor and to use that interview as the basis for a reflection on the work of being a dissertation advisor and on the ways in which both I personally and the mathematics education field more generally take on that work. The result of this work is this chapter, in which I consider Jeremy's approach to dissertation advising and engage in some questioning of how I work as a dissertation advisor.

The purpose of this chapter is twofold. On the one hand, this chapter attempts to document Jeremy Kilpatrick's style as dissertation advisor, drawing on his own reflections apropos of questions I asked on the subject. My own experiences first as student of Jeremy's and later as professor at the University of Michigan are context for the question and motivation for the interview in ways such that they bias the content of this chapter. Thus, the reader will encounter Jeremy's words about dissertation advising in that context. On the other hand, this chapter raises consideration of various contemporary issues at play in dissertation advising and the various ways in which advisors might address those in mathematics education. There is inevitably some contrast to be found between Jeremy's dissertation advising style and the styles that are desirable or even possible for other scholars including myself. The contrast may be due to differences that can be generational, institutional, or perhaps just personal. Jeremy directed dissertations during a long but still particular time period in the history of mathematics education, and at institutions where mathematics education has had a privileged place—Teachers College, Columbia University, and the University of Georgia. Those circumstances, while not determinant of a style of work, arguably enabled Jeremy to work in the way he did with individual advisees.

Images of Dissertation Work

As I came from Argentina to the United States to do a doctorate in mathematics education, I had the impression that a doctorate was an apprenticeship and initiation into a research career. Probably because at the time, many university faculty in Argentina didn't hold doctoral degrees, I had not realized that the doctorate also plays, particularly in the United States, the role of "certification" for people who would like to teach college students. This naiveté of mine regarding the role of the doctorate in the United States brought with it my assumption that individuals coming to do a Ph.D. would be disposed to do research and eager to get socialized into a field of research. It also brought with it a sense that the research apprenticeship, in general, and the dissertation, in particular, were the most important elements of the doctorate. While fields and institutions differ in that regard, it turned out that in the Department of Mathematics Education at the University of Georgia circa 1993, those assumptions were not universally shared among my peer graduate students or even our professors.

The University of Córdoba, in Argentina, where I had worked as an instructor and research fellow before coming to the United States, granted doctorates in several academic fields. I had images of what the dissertation work might look like that were tied to what I had seen friends and colleagues do in fields other than mathematics education. I had seen doctoral students in natural sciences getting apprenticed in research through membership in the laboratories where their doctoral advisors worked. These labs would form around a general focus or method but entertain a manifold of projects that new students would join in. Over time they would pick a project from those that their lab director made available to them and get a dissertation in the form of a set of articles, usually involving several co-authors that reported on the work they had done for that project. I had also seen doctoral students in mathematics study for qualifying exams first and then pick an advisor based on the field that they would be interested in working in. The advisor would give them a topic to study from within which they would develop material for the dissertation that they would share in group seminars that convened the dissertation advisor and others working on neighboring problems. The dissertation would be written up when the study of the topic had yielded some critical mass of proved results. Those images contrasted with the impression I had of students working in philosophy or history, where I noticed that doctoral students would work much more independently at the outset, choosing their own topics and working alone, and the advisor would play more the role of a consultant earlier on and that of a critical reader during the writing of the dissertation. I did not know many people who had worked on dissertations in mathematics education. From a colleague who had gone to Europe to do a dissertation based on a mathematical idea she had been working on before, I had learned that the mathematical topic she had been working on had been recast in terms of the theory her doctoral advisor was working on. An eventual dissertation topic had emerged that contained traces of her original focus but also aligned to make a contribution toward advancing the theory. Yet my colleague's own work was done virtually independently, with periodic meetings with her advisor and only theoretical connections with other doctoral students working with the professor. While those images of dissertation work were diverse, it was common to all of them the centrality of the role that the dissertation research played in those people's lives as doctoral students. Until I came to the United States, my sense was that doing a doctorate was equivalent to completing the dissertation work and had not realized how many other things (coursework in particular) could be part of what Americans call a doctoral *program*.

There is an emerging scholarship on the doctorate that documents what the doctoral programs are like in the United States (e.g., Golde et al. 2006) and also internationally (e.g., Boud and Lee 2009). Through the work of Robert Reys, there has also been an attempt to document practices in doctoral education in mathematics education (Reys and Kilpatrick 2001; Reys et al. 2007) and a published discussion about doctoral programs has gotten started. I profess no deep knowledge of this literature but have found some resonances there with the impressions I had about doctoral work.

In his essay about the doctorate in mathematics, Hyman Bass (2006, p. 107) indicates that "historically, the doctoral program in mathematics was designed to be an apprenticeship into the research practice of an academic research mathematician." Students would take courses on foundational knowledge on their first year or two and then certify that knowledge through passing a qualifying exam, to eventually framing a doctoral research project based on knowledge gained in elective courses and seminars in an area of research of interest. Similar concern about basic and broad knowledge is apparent in Breslow's (2006) essay on the doctoral program in chemistry, but he also indicates that, "in chemistry, as in other fields, the core of the Ph.D. program is research" (p. 178). And he adds:

> Generally, a student joins the research group of a faculty member and pursues a research problem that is suggested by the mentor. Also the mentor monitors progress and makes suggestions throughout the course of the program. The intensity of such monitoring varies, from 'What did you do yesterday, and what are you going to do today?' to 'Please let me know if you run into any problems or make any important observations.' ... It is quite unusual for students to propose and carry out their own research ideas, unrelated to the interests of the faculty mentor. Generally, it is a goal of the Ph.D. program to bring students to the point that they could, indeed, propose a sensible research problem. (p. 178)

In her essay about the doctorate in history, Appleby (2006) envisions a 5- to 6-year doctorate in which after 2 years of coursework covering basic scholarship, 3–4 years are spent writing the dissertation: "Writing history is what constitutes a professional historian, whether that person continues to work in an academic institution or not" (p. 324). Appleby describes this work as "completing a piece of historical scholarship under guidance" (p. 324), an experience that often requires isolation, "doing solitary, creative work" (p. 325). Appleby recommends writing groups to help each other mitigate that isolation and characterizes those groups, thus:

> Students do not need to be working in the same field to help each other as readers and critics. Indeed, the very lack of familiarity with a topic alerts each writer to the importance of clarity in the structure and writing of history. (p. 325)

Appleby (2006) also describes the role of the advisor as focused on more formal and methodological issues than on substance:

> Mentors can and should help students set parameters around their research projects, showing them how to limit their scope as much as possible so they can complete the project in a few years, with plenty of time devoted to studying the context of the event or development being explored. (p. 325)

While offering personal and brief views on the doctoral dissertation work in mathematics, chemistry, and history, the three scholars cited do help document the different roles the doctoral advisor has traditionally played in those disciplines: In particular, the chemistry advisor seems much more invested in the topic being researched by the doctoral student, while the history advisor seems more invested in the individual enculturation through the completion of a piece of historical writing. Resonating with those comments, Jones (2009, p. 30) indicates that, "doctoral education in the United States experienced a divergence between the laboratory sciences and the humanities during the mid-twentieth century, as external funding of scientific research transformed the academic workplace in many disciplines." Specifically "many doctoral students in lab sciences [are] admitted to perform a specific laboratory analysis funded by a larger project directed by their primary advisor.... These students generally receive their Ph.D. when the lab analysis is complete" (p. 30). In the humanities, in contrast, "students [are] expected to select and design their own research and spend many years working alone on their projects" (p. 31). Reporting on how scholars have reacted to the differences among their fields' traditions of dissertation advising, Jones quotes: "What do you mean, the student is assigned a research topic?' from humanities faculty and students, and 'Do you really think it's responsible to expect your students to spend 10 years in your program?' from the science side" (p. 30). Jones notes that each tradition has strengths and weaknesses:

> The strengths of the lab science model at its best are: frequent contact between student and advisor (daily-weekly), the immersion of the student into a lab group that forms the basis for peer mentoring and a sense of intellectual community. (p. 30)

But among its weaknesses, she notes "few opportunities for students to develop independence and creativity" (p. 31). In the humanities, Jones notes as strengths "the expectation of independence ... and the flexibility to choose and change topics" (p. 31) and as weaknesses "infrequent contact between student and advisor [and] isolation of students from peers and advisors" (p. 31).

Is there a signature form of dissertation advising in mathematics education? The report by Reys and Kilpatrick (2001) includes very little about the nature of the dissertation work. In their chapter on the research preparation of doctoral students, Lester and Carpenter (2001) include a paragraph on the role of the dissertation. Lester and Carpenter (2001) note:

> Before about 1980 the dissertation was begun at the end of a student's coursework and often was viewed as the final hurdle to be cleared before earning an academic "union card." Today, those doctoral programs that emphasize the preparation of researchers regard the dissertation as a culminating research activity, following three or more years of research-related experiences. (p. 65)

And they also note that, "the role and importance of the dissertation in the research preparation for doctoral students vary from institution to institution" (p. 65).

But the majority of the Reys and Kilpatrick (2001) report is dedicated to discussing aspects of the doctoral curriculum in mathematics education other than the dissertation, especially the coursework in mathematics and in mathematics education that doctoral students take. Courses were a prominent part of my experience as a doctoral student at the University of Georgia as well. They contributed in important ways to my socialization into the field by complementing the less systematic reading I had done before starting the program and the more focused reading I eventually did for my dissertation work. The need for core coursework in mathematics education is quite apparent when one realizes that the doctorate in mathematics education in the United States does not have an academic undergraduate or master's program feeding it. Students who come to a doctorate in mathematics education have often had undergraduate degrees in mathematics or in mathematics teacher education that rarely have given them any acquaintance with scholarship in mathematics education. Yet while coursework provides opportunities to acquire the knowledge produced in the field, they rarely provide apprenticeship into producing such knowledge. The creation of opportunities for research apprenticeship has been an important issue for me during my first 10 years as a faculty member at the University of Michigan, and it relates both to the images of doctoral work I had from other fields and to my own experience at Georgia.

My Own Experience Writing a Dissertation Under Jeremy Kilpatrick

I came to the United States from Argentina, with a Fulbright scholarship to do a master's degree and the intention to continue on for a Ph.D. I had originally applied to the University of Georgia because Jeremy had been suggested as a doctoral advisor. Immediately prior to that, I had been working as an instructor and research fellow in a mathematics department that gave doctorates in mathematics. And I was a member of a mathematics education research group in which I had read quite a bit of mathematics education scholarship. Having read theory and research with my group, I probably had more knowledge of scholarship in mathematics education than other incoming graduate students before I started my studies at Georgia. Yet that knowledge was not only very partial but also disorganized. It is clear to me that the extensive coursework I took in Georgia helped fill gaps in my knowledge of mathematics education and mathematics. In that sense, I could see the value of a "program" that included some required coursework. However, I had mixed feelings about the extent to which those aspects of the program took time and some prominence away from research apprenticeship.

On the one hand, the way coursework and research apprenticeship were combined was not completely satisfactory to me. The dissertation project started only after the student had passed the preliminary examinations, after the completion of coursework. There might be a reason for it in that probably only at that time the student could conceive and defend a prospectus of a study, and dissertations were, for the most part, conceived as individual, student-run projects. I had experiences in research before my dissertation, through informal and paid involvement in other professors' research projects, but these experiences did not create any programmatic continuity with the projects I did first for a master's thesis and later for a doctoral dissertation. Both of these projects I developed independently from any work I had been involved in. The various assistantship positions in which I was involved provided me with skills that I could reuse in later projects, but they did not help develop a sense of research program.

On the other hand, working independently for the masters' thesis and the dissertation was empowering. As advisor of those projects, Jeremy was quite open to my bringing in of literature that I judged relevant to frame each of the studies, even if he was not familiar with it. He was open in letting me develop research questions, research design, and the general orientation of each of the projects. Jeremy was also quite forthcoming in providing suggestions that contributed to the line of thinking I was pursuing, including, in particular, literature that I had ignored. Most importantly, Jeremy was extremely generous with his time and knowledge as he interacted with me through my writing. Jeremy would take multiple drafts of chapters and work through them in detail, getting at how the ideas were constructed through language. Jeremy's reactions were developmental: He would often correct the grammar or word choice, but he would also provide a rationale for his correction in ways that slowly helped build my knowledge of language and my skill at writing. I had, and probably still have, a process view of writing in which writing is a tool to help me think through issues: Along those lines, I write with the main purpose to inform myself. This writing style was (and probably continues to be) well adapted to, following my train of thought, not shying away from digressions if they seemed tactically appropriate, and not providing enough clues to help the reader capture the structure of the argument or to understand the strategic role of what they were reading. Jeremy was generous not only in providing feedback on what I wrote but also in insisting that I streamline arguments and help the reader know what he or she was reading. His summative reactions always articulated the position of the reader: "The reader does not need to know your intellectual history," he told me once. Rather than taking it easy on me by trying to understand what I might have meant in my writing, he focused on rendering for me the readings entailed by what I had actually written, which quite often prompted me to improve or rewrite. Only occasionally was his summative feedback nondevelopmental (e.g., "this does not make any sense").

In hindsight, it seems as though having me write about something that I had played the main role in developing was instrumental for that interaction about writing to work: As I wrote, I was driven by the need to make myself understood. And as I revised, I was driven by the need to explain my ideas to Jeremy, who took care to represent very well the position from which other readers would be coming. The commitment I had to what I wanted to say was important for me to mitigate the frustration I felt and to fuel the effort I had to make in ways that I can appreciate only now: Had I been writing about ideas that I did not play such a role in shaping, I might have been less invested in the need to get the writing right and less forgiving of Jeremy's focus on what the text said rather than what I wanted it to mean. It is an understatement that Jeremy was my first and most dedicated writing teacher. He was also my first and most dedicated reader. The amount of work and patience that Jeremy put in is better appraised when one also considers that I had had little writing instruction even in my native Castilian: My formal instruction in writing had consisted mostly of work on sentence grammar and word choice, and little to no instruction on composition, and I had also had very little academic writing practice in my undergraduate college experience.

In my years of mentoring graduate students through scholarly papers and dissertations, I have often recognized Jeremy's influence in my attempts to relay what the reader's experience is when I respond to students' writing, and I realize how long a way I've come. As I write this, it becomes apparent how instrumental it was for my individual development that I had such a strong commitment to what I was writing about. It seems as though Jeremy's disposition to having me work on a topic of my choosing, letting me frame and design the dissertation work, was an important component to make possible the development of such individual writing capacity.

In my own career as a university professor, I have not always seen individual development as such an important a part of the doctorate. My efforts as advisor have been more oriented toward making the doctoral experience one of socialization into the field through immersion in a research program that contributes to scholarship in the field. I particularly like the laboratory science model of research apprenticeship and have had local examples in the School of Education at Michigan that showed what that could look like. The fact that in Michigan I was expected to develop my own career programmatically also put a premium on developing my doctoral education efforts around concurrent themes of research. And since the structures we have for funding doctoral students are largely dependent on externally funded individual faculty projects making large expenditures for tuition, stipend, and benefits for research assistantships, I was encouraged not only to pursue research funding that enabled me to have graduate students but also to see those research assistantships as paying both wages for work and commitment, on the part of the doctoral student, to contribute to the line of scholarship funded by the project. After a few years of trying to enact a laboratory model for doctoral preparation with various degrees of success, I've kept questioning whether that is a fitting style for mentoring dissertation work and, coming back to my experience studying with Jeremy, wondering about what I've missed in my understanding of what the dissertation work is about that might prompt a redesign of the way I carry on that work. I am also at a stage in my career when I can afford less tunnel vision, and, at least conceptually, dedicating more time to read work whose substance I may not personally be invested in. The idea of interviewing Jeremy about his style of dissertation advising serves this questioning process of mine in addition to documenting for the field how Jeremy perceives his work as dissertation advisor.

Jeremy Kilpatrick's Acknowledged Style of Dissertation Advising

In December 2011, I spoke with Jeremy about dissertation advising. I asked him to describe his style as director of a dissertation. Jeremy answered:

> I don't know if *style* is the right word, but what I do try to do is—.It's the student's dissertation, and so in my view it's the student's responsibility to come up with a problem and to figure out how to investigate that problem. And I view my job as trying to help the student. To offer [students] reasonable feedback on what they propose to do, trying to help them formulate their research question, develop some sort of framework for studying that question, and

figure out … what methods they are going to use to get data that would speak to that question. But in my view, it is their project. It's a part of their developing as a scholar to be able to formulate and carry through a dissertation question. Well, multiple questions, actually. But I call it a question because they need to have some kind of issue that is driving what they do.

Most students have not done this sort of thing before, unless they did a master's thesis. So, for most students, … this is the first time they've done it, and it's a very difficult thing to figure out how to do. So I try to be as helpful as I can without telling them what to do. I know that there are a lot of different ways to do this, but I … I tend to keep hands off if they want to go in one direction…. If I think the direction is not going to be particularly productive, I might say so. But I will let them pursue whatever they want to pursue as long as it looks to me like it's going to result in some kind of a productive study that contributes to the field. But for me the primary issue is, does it contribute to their development as a fledgling scholar?

Jeremy's response to that first question gives us a sense of how he views the dissertation work: According to Jeremy, the dissertation work is the student's project, and its main purpose is the student's development as a scholar. Rather than "telling [the student] what to do," Jeremy conceives his role as keeping "hands off if the student wants to go in one direction." His role does include helping students translate the "issue that is driving what they do" into a dissertation question; it includes helping the student develop a framework to study the question and identify methods to get data that speak to that question. Jeremy elaborates on how he does that thus:

I figure it's partly my job to try to suggest things that they go and read that are connected to what they want to do. In other words, I need from them a very cursory statement of what they're planning, … of what they are thinking about. It can either be written down, or they can tell it to me. And then I can try to find some reasonable references that they can look at in order to get into the literature in that area. Of course, that always depends on me being able to find those things myself. But now with the Internet, I can search for things that might be helpful, as well as recall things that I know about and loan them books and suggest directions for them to go in. So the feedback that I give is usually in the nature of, "Have you looked at this?" "Here's an interesting study that you might want to look at and think about." And so on. And so that's, … that's really how I help the student get into a particular domain for research. I assume [students] will change their research question, but… in most cases, they do not change the general ballpark that they are looking in. So I can usually find some things in that ballpark for them to become familiar with, and that helps them develop their sense of what the field has done with that issue.

Interested in knowing whether that approach had boundaries, I asked how he would respond to a student whose topic of interest he perceived as too idiosyncratic or far off the range of things that mathematics educators investigate. Jeremy noted:

Let me give you an example that comes from my experience at Teachers College. It came at the end of my time there, and so I didn't get the issue resolved with the student. But a student came to me and said, "I would like to study the connection between astrology and mathematics learning." She didn't mean just what the newspaper prints or what Zodiac sign you were born under. She really meant, where were the planets when you were born? And what phase was the moon in? And all of that. It was a very elaborate view of astrology. And she wanted to know … how that might be connected to whether people were good in mathematics or not. As you can imagine, I was very much taken aback because it seems [to me] that there's essentially no room in our field for studies of astrology. But I talked with her and got her to try and consider how she might shape this into something reasonable to do. I never was able to complete that, and I frankly don't know … if she finished a degree at Teachers College, and if she did finish, whether she studied astrology or not.

But that strikes me as one of the most, ... probably the most, idiosyncratic choice of topic that I can think of. Because there isn't—. ... To my knowledge, there is nothing in our field that connects astrology with being able to do mathematics. And I couldn't see how you would turn that into a researchable problem. But I didn't immediately say, "This isn"t going to work." I let her know that I didn't think it would work very well, but I wanted her to keep thinking about it and see if she could turn it into something useful. Because, who knows? She might have been able to reconfigure it into some, ... some reasonable kind of problem that I couldn't see from where I was sitting.

So that's an example of [how I handle] idiosyncrasy. If [students] have some bizarre topic that they want to pursue, I let them go as far as they can with that until they convince me and the members of their committee that this is something worth doing. I don't think this student would ever have been able to convince the whole committee that this was something [reasonable] to do, but who knows? She had just started working on the problem, and she hadn't really thought much about it, and I was going to let her see what she could come up with.

Following that approach is, according to Jeremy, consistent with the notion that the dissertation is the student's project and its main goal is the student's development as a scholar:

Well, again, I think we need to look at the dissertation as an exercise in the student's development. So the student is doing research for the first time, probably. And it's [the student's] job—. ... I feel the formulation of the problem is as ... important [as] the way in which the problem is addressed. So I figure students ... *may* need—. ... They usually *do* need to struggle a little bit with what exactly is [their] research question. [They] need to struggle with it, ... reformulate it, ... think about it, and ... look into the literature to see if there's anything, ... to see what has been done that connects with this problem. And then, [they] can go with that problem.

This is actually the way my own advisor worked with me, Ed Begle. This is the way he let me do what I—. ... Pretty much what I wanted to do. And [he] did very little shaping of my problem. Because I think he thought as I think: that the dissertation exercise is something that the candidate should pursue on his or her own—of course, with support and guidance from a major professor and a committee. But it's the student's problem. It's something that is helping the student enter the field. And I think it's a mistake for us to try to hand the student a topic or insist that a student go in a particular direction.

Considering that many mathematics educators work in departments that gather more than mathematics educators (e.g., curriculum and instruction), I asked Jeremy how he would respond to a student whose question of interest was taking him adrift from mathematics education, either to an education question in which the subject matter was completely generic or to a question that had nothing to do with education. It is important to note that while Jeremy did his Ph.D. at Stanford's School of Education (which is a place where mathematics educators are located in the same academic unit as education researchers working on many other fields), his advisor (Ed Begle) and one member of his committee (George Pólya) were mathematicians. Also, Jeremy has spent all his professional life (at Teachers College and the University of Georgia) in departments where mathematics education has had quite a bit of independence. Jeremy responded:

I try to make a point with anybody that I take on as a student ... that we are in mathematics education, and therefore whatever problem they come up with to pursue as a dissertation topic has to be a topic in mathematics education. And that means, in part, to me that mathematics cannot be treated as a black box, as something unexamined. [It] happens a lot in fields like sociology or psychology that mathematics is simply taken as something not to be explored. Whenever I work with [students], I say to them that their problem needs to take on mathematics as problematic, as something that needs to be opened up and explored, not as something—as I say—as a black box. ... That's the metaphor that I use. Don't treat mathematics as something we all—. ... We all know what that is, and therefore I won't look at it in my study. I think they have to somehow engage with the question: What does mathematics mean in this study? And what, ... how can I bring my knowledge of mathematics to bear on what's going on in this study?

I also expect them to draw upon their experience as a teacher, as an educator, in the same way. It has to be a study in education. That's where we are located, and that's what our field is: It's mathematics education. What that should mean is that the dissertation that they have done is actually a contribution to the field of mathematics education and not just sociology, anthropology, psychology, linguistics, or something else. It can touch on those fields, but it is a mathematics education study, in my view, or it should not be done in this department with me. They can use literature from other fields in developing their theoretical framework, in trying to figure out how they are going to do the research, but at some point they have to consider the mathematics in what they are doing and take it as problematic or as discussable. I don't particularly care how they draw on the literature for that as long as they understand that school mathematics is not something unproblematic. If they are looking in schools, school mathematics is not something unproblematic; it's not something that everybody understands the same way. [They should understand] that trying to give a portrayal of what they mean by *mathematics* in their study is part of the job. It's not necessarily connected—. ... It's not necessarily the main literature that they draw on. They may be drawing on work in, let's say, sociology for their main theoretical ideas. That's okay. But at some point, they've got to make contact with mathematics, or it's not a study that I would direct.

I asked Jeremy whether he had had to handle students who wanted to work on a research question in which mathematics was only an element of context. He said:

I can't recall an occasion [like that] where anybody came to me to be the major professor. Now I do recall occasions where I was on the committee, and the problem that the student brought to the committee didn't seem to have much to do with mathematics. And if that was the case—. ... Of course, if it was in another department, then I didn't worry about it. But if it was in my own department and didn't seem to make much contact with mathematics and mathematics education, then I made my opinion known there that the student should try to move the study a little bit more in the direction of what's mathematical about this. You can see this [issue] sometimes in the work people do. We've had presentations in our department where people presented their work, and ... a question that gets asked is, "What's mathematical about what you're telling me?" And sometimes it's a hard thing for people to answer.

It appears from the previous comments that while Jeremy's acknowledged advising style is such that the student plays a central role in defining the dissertation topic, the student's activity is done in a context and with elements of support that serve to regulate that liberty. In particular, the academic units where Jeremy has worked, both at Teachers College Columbia University and at the University of Georgia, have allowed some amount of independence to mathematics education. Mathematics Education was in an independent department at Teachers College before 1975. Mathematics Education was also an independent department at the University of Georgia when Jeremy joined in 1975 and until 2004, when the

College's administration required that it merge with science education. But even then, mathematics education has maintained quite a bit of independence as a doctoral program, and the same can be said about the program in mathematics education at Teachers College. Also, while in Jeremy's view, the student is responsible for coming up with a topic for research, his own responsibility as advisor is that of supporting that process by bringing both challenge and resources. Jeremy pushes students to connect their interests to others' work in the field as well as to make apparent how the topic examines mathematics as well as education, and he takes it upon himself to bring in bibliographic resources that the student may not have come across but that may be relevant to the student's work. This was certainly true in my case.

I personally recall some of how that happened in my own case. Jeremy has always wanted students to read other people's dissertations (rather than just the articles based on the dissertations). When I started my dissertation work looking at what acts as proof in secondary school mathematics, Jeremy pointed me to Yasuhiro Sekiguchi's (1991) dissertation on proof in high school geometry, written also under Jeremy's direction at the University of Georgia, and one that I had not come across in any of my literature searches because it had not become an article. While at the time it appeared that Sekiguchi's dissertation focus was much narrower than what I intended to write on, the recommendation was quite important in helping me realize how salient the two-column form was in the experience that US students have doing proofs in the geometry class. My dissertation looked at proofs and other justification strategies in high school mathematics more generally, but my career afterward zoomed into two-column proofs and the geometry class in ways that were greatly facilitated by my reading of Sekiguchi's dissertation. In particular, Sekiguchi's review of the literature on the teaching and learning of mathematical proof contributed to my socialization into the field of mathematics education. This is important to bring in here to complement Jeremy's description of the dissertation as a student's project: While the dissertations he directs may respond openly to the intellectual questioning of the student, the resources and challenges he provides, either directly with his questions and reactions or indirectly through the readings he recommends, contribute to make the dissertation responsive to the corpus of prior writing and the ways of thinking and writing that are customary in mathematics education. The fact that Jeremy has done this work in the institutional context of strong mathematics education doctoral programs located most of their time in mathematics education departments has likely helped make that indirect influence effective.

The Dissertation and Its Stakeholders

I would like to propose a way of thinking of the dissertation as an outcome that responds to the influence of stakeholders. Jeremy's comments about his advising style clearly identify the student as a main stakeholder. But as he elaborates on how he helps the student, the field of mathematics education appears also as a stakeholder: The topic the student works on needs to relate to literature in the field that Jeremy helps identify, the topic needs to examine the mathematics, the topic needs to relate to teaching and learning, and so on. I would like to argue, however, that other stakeholders are present as well. In particular, as I note above, I suspect that for Jeremy it has been relatively easy to represent the field of mathematics education in his advising role because of his institutional location in mathematics education (in either independent departments or independent programs). Other advisors whose appointments are in institutions less directly identified with the mathematics education field may have more difficulty handling their role as steward of the field of mathematics education and faculty member of an institution.

As a transition, and to flesh out how different institutional conditions may be and how these differences may shape the dissertation, let me note how different the situation can be by describing my own position at the University of Michigan School of Education. The position I took in 1999 was earmarked as secondary mathematics education because I was expected to take responsibility for the teaching of a secondary "methods" class for prospective teachers of mathematics in Grades 7 through 12. The doctoral program, however, was and continues to be a doctoral program in Educational Studies, where mathematics education is one among several areas of specialization. The school has no departments; instead, the faculty is grouped around two programs (Educational Studies and the Center for the Study of Higher and Postsecondary Education). The mathematics education area of specialization is one among three areas of specialization in the "education in mathematics, science, and technology" (EMST) unit within the Educational Studies program. While applications to the doctoral program in Educational Studies require applicants to choose an area of specialization and mathematics education faculty recommend students' admission to our area of doctoral specialization, admitted students are officially students in Educational Studies. This means, in particular, that students can switch between areas of specialization and even design an ad hoc cross-specialization. I offer that description of my own workplace conditions, which I would argue has drawbacks but also benefits compared to an independent program or department, to illustrate an alternative institutional context in which the advising of a doctoral dissertation in mathematics education could unfold. In particular, it illustrates why in thinking about stakeholders for a dissertation, it is important to think separately about the influence of the institution and the influence of the field.

Four Stakeholders

I would like to propose that there are four stakeholders of the dissertation: (1) the student, (2) the field of mathematics education, (3) the institution where the dissertation is submitted, and (4) the advisor. Each of those stakeholders could accomplish things through the completion of the dissertation. In an exercise of speculation, I examine those stakeholders in terms of how they could conceivably be invested in the successful completion of a doctoral dissertation. In the next section, I show what Jeremy had to say on the matter.

The student is clearly a stakeholder in the dissertation. Students could and actually do think of their dissertation in many different ways, probably derivative of how they think about the doctorate in general. At one end of the spectrum, there are doctoral students who see the doctorate as a process of apprenticeship into scholarship in a field of study and thus see the dissertation as one project in which they can learn to author work that contributes to that field. Those students may thus expect that the dissertation will bring them closer to belonging in the circle of producers of knowledge that has public value at least for other workers in the field. In an extreme, this approach can be alienating or can involve the student in work that, while important for the field, has no personal meaning to the individual. At the other end of the spectrum, there are doctoral students who see the doctorate as a credential or an attestation of individual worth and the dissertation as an individual challenge by which to prove themselves. They may expect independence in defining a topic and in coming up with the resources to pursue it and write about it, putting a high premium on studying what they are personally interested in, with the theories that they personally see as compelling, and the methods that they see themselves as skillful with, finding their own voice through the writing, and being done with the project in a timeline that they can control at least partially. In the extreme, the dissertation may have no value for the world of knowledge, but have value for the student as an accomplishment in writing or as the symbol of reaching a high stage in individual development. The two goals of apprenticeship and individual growth are more or less combined in every dissertation, but they may participate in the mix in different amounts. Thus, the way in which the student is a stakeholder of the dissertation may combine in different ways the use of the dissertation as a means of socialization into the ranks of productive scholars in the field and the use of the dissertation as a means of personal expression and establishment of individual worth.

The field of mathematics education is invested in a dissertation in substantial and in formal ways. Since dissertations use a large amount of the energy and resources available for research in the field, the substance they are concerned with matters in regard to whether and how well our field builds its theoretical and methodological basis, how rigorous our field's knowledge is, how fast our field's knowledge grows, and in which directions it grows. In a field so young as mathematics education is and one which continues to be perceived publicly as a field of applications of theory and methods from other disciplines (which conditions, e.g., what sorts of research in mathematics education are funded), it would be of interest to the field if dissertations were seen as opportunities to contribute to the investigation of questions that are strategic for the development of the field (pace the pull from the student stakeholder side). But even if the field did not make strategic use of the substance of dissertations for bolstering its presence in the concert of academic knowledge, dissertations do matter formally. Dissertations matter to the field because the formal consequences of a doctorate impact the actual conditions of existence of the field: Individuals with doctorates in mathematics education form the pool of people who aspire to faculty positions slated for mathematics education scholars. These positions, in turn, are, as of now, the main places from which mathematics education scholarship is done: These new faculty members will have to engage in research production to maintain their positions; they will be getting research funding to support that scholarship; and, if those positions exist in doctoral granting institutions, they will be providing guidance for new scholars. The dissertation is a keystone in the individual scholar's insertion into the field: It can be critical in getting their research career started—the new scholar is likely to develop his or her scholarship from the dissertation, both insofar as initial publications as faculty member will derive from the dissertation and because new topics of research may spring from or at least relate to the dissertation topic. Therefore, to the extent that positions in academia are instrumental, in the manner described above, in the development of fields of scholarship, each dissertation has the potential to contribute to define what the field as a whole ends up being occupied with. This contribution may or may not be a positive influence in the development of the field: The field is invested in each dissertation in the sense that it can gain or lose from it. Of course, it is not necessarily the case that the field will lose if the doctoral candidate works on a topic of no present importance for the field. One could also understand the field's investment as one on human resource development whereby supporting some autonomy for the fledgling scholar (e.g., in the choice of topic) may be a reasonable compromise in exchange for imposing some standards of rigor (e.g., in the solidity of the framing or the research design). Allowing some autonomy might be instrumental in such human resource development goal in that it could hold the new scholar accountable for all aspects of scholarly work including justifying the importance of a dissertation topic or managing the logistics of data collection, to name a couple. These experiences during the dissertation can serve as a trial run on work that the scholar will soon have to do on his or her own. Autonomy in the choice of the topic may not only help the scholar learn from experience but also, conceivably, draw the most energy from the scholar. Furthermore, it can also enhance the chances for the field to open new scholarly paths, drawing on the new scholar's originality and drive to open new questions or new ways of addressing questions. And autonomy may sustain the new scholar through the ups and downs of dissertation work, in that steering the dissertation to completeness may feel more like his or her own self is at stake.

At the complete opposite extreme of autonomy in regard to the field's benefit is a position whereby the field might benefit from the new scholar's circumspection in regard to his or her own interests, letting the dissertation advisor as a representative of the field take more responsibility for the orientation of the dissertation work by identifying a research topic and general question that the field considers important to address. This reduction in the autonomy of the beginning scholar on behalf of the field may be instrumental to promoting the development of critical mass of studies in areas of importance and exploiting existing resources such as existing data archives or new theories being explored. While more imposing, this approach has the potential to enable the field to grow in its capacity to answer important questions, as well as to ensure that the scholar's initial contributions (hence, his or her initial steps in a research career) will be relevant to the field. This, in turn, can benefit not only the field but also the scholar in that his or her scholarship may be more likely to be of interest to reviewers, journal editors, book publishers, and readers when seeking publication. While the scholar is likely to grow in autonomy after the dissertation, his or her development will proceed from a starting point that the field can own up to.

The institution (i.e., the program, department, school, and university) that confers the doctoral degree is also a stakeholder of the dissertation. Some of it is apparent though it may seem completely formal. For example, through their graduate schools, universities specify, approve, and regulate doctoral programs. They regulate the steps that doctoral students need to take before they can officially submit a dissertation for consideration, establishing requirements for candidacy and devolving some power to faculties to enforce those requirements. They determine who can direct dissertations and how the dissertation committees need to be formed. They regulate and enforce writing formats; they establish deadlines and defense protocols. At the very least through the imposition of those requirements, universities and their graduate schools remind doctoral students that the dissertation is being done within an institution and that the dissertation has to be accountable to the institution. Those regulations, in spite of their formality, are meant to achieve some basic standards of quality and comparability among the dissertations submitted to the university. A university needs to be able to stand behind the degrees that it confers, and the formal standards that they impose help achieve that goal: Dissertations can be endorsed as meeting some standards that are perceived as likely to go along with acceptable scholarly work.

The college, school, department, or program where the dissertation is developed also contributes affordances and constraints to the process. These can be of the same formal nature as those of the graduate school, but they can also be more ingrained in institutional culture. A dissertation presented to the faculty of a school or college of education, where the doctoral committee may include other faculty than mathematics educators, might require, for example, a problem framing that relates to education policy or practice writ large; theory, design, and method considerations that relate to the larger fields of social and education research; and the writing of conclusions that potentially relate what has been done to the teaching and learning of disciplines other than mathematics. These requirements need not be explicit and yet they may be practiced as if they were the norm. Those tacit expectations influence the nature of the dissertation, making it respond to intellectual and political concerns more or less salient for the culture of the institution at large.

Finally, I argue that the dissertation advisor is also a stakeholder. Dissertation advisors are also developing a career, not only as higher education faculty members but also as researchers in the field of mathematics education. The actual work of advising doctoral students ordinarily counts as part of a faculty member's teaching work: For example, while writing their dissertation, advisees often register for independent study credits that generate tuition revenue. But the whole package of deciding to admit or fund a doctoral student, accepting to direct dissertation work, and dedicating time to the dissertation advising has to, I argue, be recognized as relying on more than teaching responsibility.

The dissertation advisor is indeed a stakeholder insofar as he or she is a teacher of higher education. In terms of this identity, one could further examine the advisor as teacher of a doctoral student (mentor) or teacher of a field of scholarship (steward of the field). None of these teacher roles necessarily involves the faculty member's own research work in their stakeholder role and yet they suggest different ways of being a dissertation advisor. At one extreme, the dissertation advisor may see himself or herself as a mentor of a doctoral student and consider himself or herself personally invested in executing a style of mentoring—a way of bringing individual students up from whatever stage in their development they are at and up to the next stage. At the other extreme, the dissertation advisor may see himself or herself as a steward or gatekeeper of the field of mathematics education, personally invested in the communication to the student of a knowledge canon and way of working that are characteristic of the field. One could assimilate these dual considerations of the role of the faculty advisor to the role that any teacher has to play in classroom teaching: teaching their students as well as teaching their subject matter. But while those considerations may already make the faculty member's involvement in dissertation advising complicated, there are further complications that arise from considering that a faculty member is not only a higher education teacher.

Unlike K-12 teachers, higher education faculty members are expected to actively produce public knowledge in their field by pursuing a research agenda, especially in doctorate granting institutions. This work is not only part of what they are expected to do but also, quite often, the main reason that individuals pursue careers in academia. A position as professor provides some elements of infrastructure (autonomy, connections, facilities, library, professional staff, time) with which an individual scholar can define and pursue a research agenda. Those resources are harder to secure for those working independently or in a different kind of institution (e.g., a nonprofit). One could take a principled perspective and assert that dissertation advising is just part of the teaching load that a faculty member is obligated to take on and that it should be separate from the faculty member's own scholarship. I'd contend that looking at dissertation advising only as teaching would fail to account for differences in the manner in which different advisors advise or an advisor advises different doctoral students. Instead, I claim that the research agenda of the individual faculty member can also be at stake or that individual faculty members, to the extent that they remain active as researchers, are stakeholders of a dissertation also insofar as researchers.

The faculty member's management of resources such as funding or time is a place where to note how they are a stakeholder of a dissertation also insofar as researcher. An appealing reason that scholars want to do research as professors in universities is the access they have to graduate students. Graduate students who get involved in a professor's research, unlike employees in a research lab, are likely to be involved in the work not just to earn a living but also to learn. While they may cost more money to support and take more attention and patience than paid employees, graduate students have appeal in that they may carry forward what they learn through their research apprenticeship into their own work when they finish their studies. How could these considerations matter in shaping the way a dissertation advisor is a stakeholder in the dissertation? On the one hand, he or she could take the position that whatever graduate students do for their own doctoral research should be kept separate from the work they do as graduate research assistants in the lab or project where they work. That option may still allow for the graduate student to reuse ideas and methods learned in their apprenticeship yet require graduate students to produce their own design and human subjects protocol, collect their own data, and run their own analyses of such data, in addition to writing the dissertation. This option may appeal to some faculty members in that it simplifies matters: For example, when doctoral students graduate and leave, it would not be necessary to worry about how to manage their access to data in the custody of the faculty member's research group. Also, authorship in publications may be easier to discern if all decision-making and key roles were played by the graduate student doing the dissertation rather than shared with other members of the lab. On the other hand, the dissertation could also be carved within the scope of work of the professor's research group. While doing that may be more complicated (e.g., it may require better management of personnel, time, and access after the dissertation is completed), it has appeal for the professor: Having access to data corpuses, intellectual resources, and material resources from the group, the graduate student is more likely to contribute powerful work to the line of research of the group and increase visibility of the research group through publications. The use of faculty time may in fact be different in one and the other case. It is predictable that the students' project may receive more attention and more time when it is carved within the scope of work of the professor's research group than when it is run independently of it. But again, this may be a mixed blessing.

Thus, the considerations above describe four stakeholders of the dissertation: the student, the field of mathematics education, the institution where the dissertation is done, and the professor who serves as dissertation advisor. In each case, the ways of being a stakeholder have been explicated by characterizing polar opposites. In the case of the advisor, I characterized two sets of polar opposites: one as teacher and one as researcher.

Jeremy's Reaction to the Four Stakeholders

In my interview with Jeremy, I asked whether and how he saw those four elements—the student, the field, the institution, and the advisor—as stakeholders of the dissertation. Of course my question to Jeremy was much less elaborated than the explanation above, but our conversation did touch on the issues covered by that conceptualization. In the interest of representing Jeremy's own thinking on these issues, the complete text of that part of the interview follows:

Pat   Would you agree that in the completion of a dissertation there are stakeholders that include the student, the dissertation advisor, the institutions that support the work—in this case the department—and the field of mathematics education. How would you think of the dissertation work as meeting the needs of all those stakeholders?

Jeremy   Well, first of all, I would say that far and away above any other stakeholder is the student. The student is the prime stakeholder in this whole enterprise, as I said. This whole thing is really an exercise in the development in the student's competence as—and experience as—a scholar in our field. So what I see is the dissertation as a—. ... It's a project that a student undertakes in order to develop that student. Now, the second stakeholder, I would say, would be the field. The student really is obliged, I think, to do a study that has potential for impact on the field. That obviously varies considerably, but in general, I try to help the student develop a topic that the field will recognize as having some value. As far as institutions supporting work—. I suppose if the student is part of a funded project, you could argue that the project has supported the dissertation work. But in most of the work that we do ... in my case, I guess ... even if the student is supported on a project, the work that the student does on the project is separate from the work that is done for the dissertation. And in my view, it should be separate, if possible.

    And as far as my own stake in what the student does, I guess some of my ego is wrapped up in that, but not a lot. I enjoy working with students whether they do a good job or whether they do a poor job. I prefer if they do a good job, but I don't see myself as a major stakeholder in what the student does as a dissertation.

P   Okay. And I wanted to clarify something about the institution part and also ask you a little bit more on the last item you mentioned. I was thinking of the institution more broadly. The department and the university, for example, are institutions that have something at stake, at least formally, for example, in the sense that they invest scholarship resources sometimes to some of this doctoral preparation. But also they invest a diploma or a degree, which, I would surmise, might need to meet some standards that are not just bureaucratic. The department, too, and I think you spoke quite a bit about the program identity allowing you to insist, for example, on the mathematics and the education part. I hear the assumption that the students already have teaching experience, and that is the Georgia assumption. So in that sense there are sort of institutional affordances, I guess, for the dissertation that maybe one cannot make somewhere else. Like we [at Michigan] will sometimes have students who have not had [experience teaching].

J   Well, if a student comes to us without teaching experience, we try to supply that. ... Of course, it's in the interest of the University of Georgia and the college and the department that the students that come out of our program do a good job. But I view the dissertation as only one part of the doctoral program. ... The whole program that the student engages in is directed by a committee, and the student is examined on that program. And the student then does a dissertation. ... Obviously, the dissertation is in some sense the capstone of the program. But I don't think the university's or the college's or the department's reputation depends on how good that is. I want it to be as good as possible. I work to make it as good as possible. I hope the student does the same thing. ... Quality varies.... To use a metaphor, sometimes we are carrying the student across the finish line in order to help [him or her] get the degree. And we're providing a lot of help, and the student is not doing as much independently as you would ideally like. On the other hand, we have students who are wonderfully independent. They don't need to be carried across the finish line. They do a great job, and, of course, we are proud of them. But we can't take a lot of credit for that necessarily, because they may have come to us with the skills that they—. ... They already *had* the skills. We helped: We gave them a situation to make something good out of, ... and they did. But it's hard for us to claim all the credit: that we did everything that made their dissertation outstanding. They may have done that with very little ... with minimal help from us. So, the amount of help we give to a student is not necessarily connected to ... the quality of the dissertation. And therefore I don't worry too much about, you know, what does the university think of these dissertations? What does this college think of it? It's more, what's the stake that the university has in turning out good graduates? And we try to help our graduates be as good as possible, but we recognize that some are better than others.

P   In your work with doctoral committees, does this sort of institutional issue ever come up? About standards, or about "can we own up to this?"

J   No. I think what happens is that standards are internalized, and it's very difficult—. ... Sometimes, for example, I get ... at rare times, I've been asked, "Well, what's the difference between a dissertation question and a master's question?" And I'm not sure I can draw a hard line there except to say I sort of know it when I see it. A master's question is much more circumscribed, much simpler, much easier to attack, and so on. The dissertation question needs a little more elaboration. It needs some complexity. It needs, in my view, a lot of originality and thoughtfulness. But I can't tell you where the line is between those two. I just know that there is some kind of region in between what counts as a good master's thesis and what counts as a good dissertation, in my view.... The things that I would put emphasis on are contribution to the field, originality, appropriateness of the framework, what's done with the framework, [and] what's done with the question.

P   When you were describing your own stakes in the dissertation, I heard you say that there isn't much at stake for you. On the other hand, I did hear you say, "Sometimes we cannot take that much credit for the work of a student." To what extent is it important for you to use the dissertations as a way of leaving a legacy?

J   I don't think it's nearly as important for me to use the dissertation—. For me, the legacy is the student. I take great pride in the students I have had. I'm always proud when they do other things, when they go on to accomplish other things, and I am happy to say *that* was my student. It's very unusual for me to say, "Oh, and that student did a great dissertation." Of course, sometimes they have. But for me that's not the issue. The dissertation is a stepping stone along the way to the degree, and the student is actually a product not just of my work and not just of the dissertation but [also] of the program. And so I am happy to take credit, such as it is, for the students who have done a great job. But for me, it's not so much the dissertation as the student himself or herself. I am proud of my students. I don't really take particular pride in their dissertations, necessarily. There have been some that I have pointed out to people as good examples for somebody to follow, but that's not necessarily how good I think the student is. Because it's unusual to be a very good student and do a not-very-good dissertation, but it does happen sometimes.

P   So, what about … in terms of the topic, Jeremy?

J   The issue of topic? Well, I'm particularly happy when [students are] able to develop a topic that becomes a stepping stone for the rest of their career. They don't need to be tied to that topic. In fact, they should develop it further and go in whatever direction after that that they think is reasonable. But when their dissertation serves as the launching pad for what else they do in scholarship, I am very pleased with that. That, for me, means that the student has had good preparation. But it's not *just* the dissertation. It's in everything the student did in the program.

P   And do I hear you correctly that regardless of whether the topic is connected to what you have done personally, you would feel just as proud of the student?

J   If the student has done something very close to what I like, what I do, what I think about, then I've had opportunities to coauthor with those students in other venues. But, if the student is working in an area—let's say anxiety, math anxiety, which is an area that I don't do any work in—that's okay. If that's what they want to do, and they want to work with me, I am happy to work with them and to try to help them develop whatever they want to do in math anxiety as far as they want to take it.

P   Do you feel anything in particular about the student not being willing to use what you know or your expertise?

J   Well, if the student doesn't want to take my recommendations, that's okay. If [students] can … if they have other ideas and can defend those ideas, that's great! That's what scholarship is all about. They don't have to copy me. They don't have to agree with me. I try to be helpful to them, but they could work in a different area, and I could say, "Well, I don't agree with that. I don't believe in that, but I will support you if you can defend what you are doing." And that's fine! Our field should be able to tolerate people who have gone off in different directions. And I try not to be directive. If they need some help, I am happy to provide it. But they don't have to agree with me, and they don't have to follow in the direction that I point them.

P   So the connection seems to be more at the personal level than at the level
        of the actual scholarship?
J   Yes, that's right. Many of my students do not pursue topics that I person-
        ally would pursue. But if I think they have potential, and if I can help
        them, I will.

Thus, Jeremy's response to the stakeholders' question illustrates one possible
compromise position for the dissertation to exist within and among the four stakeholders: The student and the field are paramount in Jeremy's considerations of what
the dissertation needs to accomplish; they are the major stakeholders. In Jeremy's
personal circumstances, the department and program have protected him from having to also contend with institutional demands foreign to mathematics education
that might stake claims on the dissertation to the point that these are not too visible
for him. Jeremy does acknowledge his stakeholder role as an advisor mostly around
his responsibilities as a teacher. Students contribute to Jeremy's academic legacy in
terms of their eventual participation in the field regardless of what they do their
research on.

Other Ways of Being a Dissertation Advisor

There are, conceivably, other ways in which the dissertation work may satisfy the
demands of its various stakeholders. Those may imply different ways of being a
dissertation advisor. I asked Jeremy his views on other ways of being a dissertation
advisor.

P   The next question that I had was around the various other models of being a
        dissertation director that are available in mathematics education. I imagine
        from our conversation, it's clear that you are aware of those.
J   I am very much aware of those, and if that's the way other people view the dis-
        sertation and want to conduct it, that's fine. My major professor, Ed Begle, had
        a very strong belief that since the federal government was paying for certain
        work that we were doing, we were to do that work, but we were not to capitalize
        on it—not to take advantage of it. At the time I did my dissertation study, Ed
        was conducting the National Longitudinal Study of Mathematical Abilities, and
        people like Jim Wilson and I and Tom Romberg were helping with that.
        I had 56 students that I interviewed while they were solving problems out loud.
        They were all part of the National Longitudinal Study, and I was able to use
        background information on them in my study. But the only way I could do
        that—. … First of all, I had to convince Begle that that was a reasonable use of
        the data. I was not carving a piece out of the longitudinal study; I was just
        making use of some background data. And it would not have worked if I were
        trying to build *on* the longitudinal study. I don't think he would have approved
        that. But he did finally approve my using of some test data and some background
        data on these students because it was subsidiary. It was in the background.

It was not the main thing, which was the interviews that I did with the students. So I was able to do that. Now that's at one extreme, I recognize. Not allowing, in a way, the student to become part of a project.

I won't mention a name, but there was a professor at an eastern university in math education when I was in graduate school. You knew [that] if you went and worked with that professor, that he had a matrix of topics that you—. ... [Topics] that were available to his students. So you did no choosing of the topic. If you chose him, you got the next cell in the matrix to work on. That spared you the problem of formulating your own topic and trying to figure out what you were going to do. It was essentially handing you your problem. As you walked in the door, you got slotted into that cell. And consequently, he was able to get students [who] filled out a whole ... matrix of interesting problems, he thought, that he had set up for them to do.

What I did was completely the opposite. I went to Ed Begle and told him what I thought I wanted to do .... We talked about it, and he finally signed off on it. But it was my idea to do it, and I did it my way. I did not, and still do not, agree with this idea that when the student walks in the door, he or she gets handed a dissertation problem and [that] what you do for your dissertation is what your major professor has decided you should do. I know that that model is sometimes used in mathematics. It's rarely, in my view, used in mathematics education. And I certainly don't subscribe to it. I think that it denies [students] the opportunity of choosing a topic that is of strong interest to them and that they have decided how to work on. And I think that's part of the doctoral program experience that anybody ought to have.... I always thought that this professor [with the matrix] was denying his students an important part of their development as young scholars.

P   I wanted to probe a little bit on that, but also ask you later to think about maybe things that are in between those two extremes. So on the matrix example, I was wondering whether you saw any pluses to it. And in particular, at this stage of the competition for jobs, for example, could the matrix approach be defended on grounds that the student will be doing something that is significant in the field that will get them published quickly? Do you see that as a plus?

J   No, I don't see that at all. It's only [a] plus if the major professor has laid out a matrix that the field would find attractive. I don't know that that's necessarily the case. The matrix that I'm thinking of was idiosyncratic to this professor. I'm not saying that the studies weren't worthwhile. They probably were. I don't recall anything about any of them, but I don't think that has much to do with being competitive on the market. In fact, I think knowing that you've done a study that you didn't formulate and [that] was formulated by somebody else, I think that weakens you as a candidate when you go out to interview. Because people will say:

> Well, you didn't choose this topic, you chose the professor. You didn't work it out. How confident can we be in your career now as a young scholar that you will be able to formulate and carry through problems of your own, given that you haven't done it so far?

P So what about models in between? The two that you presented were rather extreme, if I may [say]. Are you aware of models that blend the two ends?

J Yes. If someone is working on a research project, sometimes there are pieces of the project that the student can work on and develop into a dissertation. I personally don't pursue that model. We have people working on projects, but the work they do on the project is not connected with their dissertation. And I don't have any problem with that, as long as they are doing their work on the project. They could be contributing to another study; they could be writing that study up. But the dissertation, for me, is a separate thing. However, I recognize that there are people who would like the student to take a piece of the project study and carve that out and do the dissertation in that area. And if that's the way they want to operate, fine. I don't have any problem with that. That's their way of doing it. I guess the thing I would worry about [would be that] when they allow the student to work on the project and carve out a piece of it, how much is the carving being done by the student, and how much is it being done by the major professor? Because, again, I think an important part of this whole enterprise is coming up with your own formulation and how to address something. So, if I were operating in that way and had a project and had a student, and [if I] wanted the student to do part of the project, I would try to make it open enough that the student could do a lot of the work in figuring out: What exactly is the research question, and how am I going to address it? And then I would help the student make it part of my project.

P Do you see any pluses on operating that way?

J Yes, that I think is what funding agencies would like to see. I think they'd be happy. The student can go out and say:

> I worked on this project, I developed my dissertation out of it, and I now am connected to this project in some of the publications that will be coming out…, [I did] not just my own dissertation, but I worked on other parts of the project.

But I guess I would say, "Yeah, but you could do that without doing your dissertation in the project. It could be done either way." And … the connection to the project, I think, probably gives them a little bit of an advantage with the funding agency and with people looking at the project.

P How would you react to the [following] two statements that I could offer as pluses: If the student has a personal stake on the project work, it's more likely that they will work more on it or that they would work deeper on it, because they would be concentrating efforts. The second one is, if the student is invested in something that the advisor is invested [in], too, it's more likely that they will get more advisor time.

J Oh, well…. I don't see why it would be the case that you would give more time to somebody working on the same project you were. Maybe that would happen. But if … as I say, there's work that can be done on a project that isn't dissertation work, … I would hope that a student in a doctoral program would not only be able to work on the dissertation as a separate enterprise but also work on

[the] project. I think it's in the project that one can learn how to work with other people. You know, … people have talked for years about [having doctoral students] do joint dissertations. I'm really not interested in that sort of thing, because I think the purpose of the doctoral program is to help people develop as scholars, and much of their scholarly work is going to be on their own. I know we live in a time when people collaborate on various kinds of activities. I certainly want students to learn how to collaborate. And it's great when students can work on a project, collaborate on a project, and publish things with a group of people. But for me, the dissertation is another animal…. It's something that you do by yourself to demonstrate that you have the ability to formulate and conduct and write up a study of your own. And I think everybody should be able to do that regardless of what they are doing in a group.

P   As you see the job market change, you know, mathematics education being more competitive, and the academic expectations on people being higher in some way regarding, for example, having publications before you get a job and things like that…. Do you see [the approach that the dissertation is just a demonstration of skill on the part of the student] as possibly being a liability?

J   Well, I would hope that … before the dissertation is completed, the [students] might be able to get some publications out of it on their own, and that would be a nice thing to have happen. And it does happen. I suppose that if the dissertation were part of a project, there would be more likelihood of publications. But I don't know that that's necessarily true. I think the job market, actually, is not that bad for people who have background and experience in teaching school and for people who have a strong mathematics background, both of which we emphasize in our program. And I think the problems that people have, as I see it from some angles, is that they come out of a doctoral program and they have not done all that much in mathematics. And that's a problem. And they *may* not have had any teaching experience…. If they're coming into our department [as faculty members], we're going to expect them to go out to schools and to work with preservice teachers. And if you've never taught in a school, if you've never worked with preservice teachers, we may not be interested in hiring you even though you may have a strong mathematics background. Conversely, if you've had a lot of experience in schools and working with preservice teachers, and you don't have any math background, we're not interested in hiring you either. So, I think the real issue is not the job market, but the preparation of the people who are entering that market. And for that, I don't think the issue of the dissertation is nearly as important as the mathematics background and the pedagogical background.

P   I'm wondering how particular to Georgia are those expectations.

J   Yeah (chuckling), I guess they are! … We don't seem to have a lot of trouble putting our people in other jobs, but we do have trouble finding people qualified for the jobs we have.

So it seems that Jeremy has good reasons to enact his style of advising. His views seem to be particularly strong in seeing his work as part of a collective enterprise to form individual scholars. In this sense, the doctoral program as opposed to the doctoral dissertation seems to be the main agent and the dissertation only a component of it, where the graduate student demonstrates individual capacity. Jeremy appears to think that a well-rounded capacity for faculty work is more important than focused accomplished scholarly production in enabling young scholars to start careers.

Considering the student centeredness advocated by Jeremy along with how strongly he seems to feel about his way of advising dissertations I thought it would be pertinent to ask whether and how his style admits of customization, for example, based on the needs of particular students. I asked this next.

P   In terms of tailoring the demands or needs of a doctoral candidate…, what sorts of tailoring have you found yourself doing?

J   Well, … there are people who have a lot of trouble writing, and … I figure it's my job to help them learn to write better…. When I was prepared as a mathematics teacher, I was also prepared as an English teacher, and I did student teaching in English…. So for me, correct use and powerful use of the English language is important for anybody in our field who's going to contribute to scholarship…. I try to provide as much help as I can in editing what students write for me. Now, some students need a *lot* more help than others with that process, and I'm willing to spend a lot of time on it. But that's one of the tailoring things that I do. In other words, some people need a lot of assistance; other people don't. And if they need a lot of assistance, then I provide it if I can, as best I can. So there's an example of tailoring.

Sometimes students have trouble figuring out what the research question is, and I work hard with them to try to get that question under control. Some students, when they write the first draft of their dissertation, don't have a good sense of what it is they want to say. And I don't circulate to the rest of the committee what the student has drafted until I'm reasonably happy that the student is saying what the student wants and needs to say. So a lot of what I do in the way of tailoring comes at the end of the process when the student is writing it up and getting ready to put it before … [his or her] doctoral committee. And that's where some people need a lot of help, and some people need much less help.

P   What about the issues of the style … the [advising] style. Have you found yourself with students that might expect you … to be a different kind of dissertation advisor? For example, … students that were expecting you to give them a question or students that were expecting you to be more involved or invested in their work?

J   No, I haven't seen [that]…. I think [that] if they come to me, they know, first of all, that I won't give them a research question, and they know that it may be in an area that I'm not invested [in], let's say. *Invested* is the wrong word. After I've worked with a student for a while, I do get invested in their questions even if it's something I wouldn't be working on independently. But I don't think I've had students who—. ... I rarely have students change one topic to another. And when they do, it's because they haven't been able to formulate ... a good question. And they just simply say, "I'm going to move over to another area." But usually it's through modifications of the question that they come up with the final thing that they work on. And by that time, I'm interested in helping them, and so I'm interested in the questions even though it's not something I would work on.

P   I see. And along those lines, you mentioned that if they come to you, they know somewhat that you are not going to give them a question....

J   Yeah, I don't know [where] they learn [that]. I don't say that directly to them. But I think one of the things about our program is that it has a body of students who talk to each other, and I think there must be at least folklore that if you come to me to do a [dissertation], I'm not going to give you the question.

P   Okay. So, ... if you had to give advice to somebody like me, for example, teaching in a different program regarding ... what's the best way of handling those mutual expectations, what would you say? I mean, should I try to create the folklore about how I am as an advisor? Should I try to have a protocol for discussing that explicitly with a student?

J   I think, ... I think it depends upon how your institution is set up for having a student choose an advisor. There are—particularly ... students from outside the US— ...students who come with the expectation that they will be working with a particular person, and that's the reason they're coming into the program. And so in a way, they are choosing you in advance of ... you admitting them to the program. And that's fine! ... But I think if that's the case, then there should be some kind of correspondence as to: What is the student interested in? What ideas does the student want to pursue? And in that conversation, it should become apparent to the student—if it's *my* student—... that I'm not going to tell [students] what to do. That they are going to have to figure that out on their own....

Here at Georgia, we have—for [doctoral students] in their first year, and many of them have no idea [whom] they want to work with—we have a seminar where each of the faculty members comes in and makes a presentation. And that allows the students to become acquainted not only with the faculty member but [also] with what the faculty member is interested in. And fortunately we have a big enough faculty that the students have a lot of choices of [whom] they want to work with and what topics they want to work on. And I'm, you know, I'm not eager to get a lot of students coming to me. I'm happy when they go to other people because we need to spread the advising load around the department better than it is.... [I] allow students to come and ask if I would be their advisor. And I talk with them about what they're interested in and find out, you know, what help I can be to them. And in the case of some students, I've said, "I don't understand anything about this. Can you write me something that would help me understand? Before I say yes that I'll be your advisor, I'd like to know a little more about what it is you're interested in so that I can tell you what kind of help I might give." And if I think that there is somebody else in the department who would be a better advisor, then … I talk with the student.… I'm doing more co-advising … with another younger faculty member.… I don't have any plans to [retire], but I may be retiring soon, and therefore, the student might need to finish up with the co-advisor. But it's also nice to have co-major professors, I think. It's a good exercise to engage in, and we do that increasingly with our students.

P   So from what I hear … in regard to the way you handle mutual expectations, it seems like when the student comes and describes their interest, there is some degree of letting them know how you are going to be as an advisor.

J   Yes, that's right. And for me, … as I say, that's different for, shall we say, domestic students or students coming in without an expectation that they would work with me. They have to approach me. But on the other hand, as I said at the beginning, many students coming from overseas have made contact with me before they made contact with the department, and they expect to work with me. And therefore we need to engage in a little bit of discussion before they come.

It transpires from those comments that Jeremy's efforts to tailor his guidance to the student's needs or focus have some boundaries. Jeremy will spend more time on those who need it but will not hand [out] a research topic even if the student so wanted. In regard to whether and how Jeremy will adapt his style in order to meet mutual expectations, it seems that Jeremy does consider the possibility of encouraging the student to work with another advisor. This is a possibility partly because of the large size of the mathematics education faculty at Georgia.[1] It also seems possible because at Georgia, the funding of graduate students is managed separately from their advising.

Another Dissertation Advising Style and a Personal Learning Note

As I reflect on Jeremy's comments, I cannot avoid bringing into question my own practice as a dissertation advisor. I do believe that a lot of what faculty members can do is enabled by structures and customs in the places where they work. I took my first faculty job in 1999 at the University of Michigan. While we don't have an independent mathematics education program or a large mathematics education faculty, we have an institutional culture that accommodates relatively well the notion of a research group, project, or lab centered around the work of one or more faculty members. Starting in 2001, it has been possible for me to create a research lab, the

---

[1] At the time of the final drafting of this chapter, the Department of Mathematics and Science Education at the University of Georgia had twelve mathematics education faculty members in the tenure track.

GRIP (Geometry, Reasoning, and Instructional Practices), which had at peak moments a staff of 14, including graduate students, postdocs, research staff, and undergraduate assistants. The GRIP still exists in 2013 and houses two research associates with doctoral degrees, and seven graduate students. Several resources have helped out, including as of now four federally funded NSF grants in which I've been principal investigator and that have raised more than nine million dollars in total, fellowships from the Rackham Graduate School, and some graduate student support from the CPTM (Center for Proficiency in Teaching Mathematics), which was also an NSF capacity-building grant through the CLT program (coincidentally a collaboration between Michigan and Georgia). Beyond the resources, however, Michigan has an institutional culture that enables entrepreneurship and individual faculty independence and those features made it possible for me to build up a research group around my own research program, of course at the cost of dedicating in the meantime some serious amount of time to fundraising, project management, and human resource management. An institutional culture that requires programmatic and groundbreaking research from its faculty members encouraged me to focus my efforts on my own research agenda rather than follow multiple threads. As a result, I had the opportunity not only to pursue my own research and writing but also to create a surplus of resources for more research—in particular this included the development of corpuses of rich media data.[2] Those corpuses have enabled graduate students to develop their own investigations toward degree milestones and the dissertation. In no case, have I handed out ready-made research questions, but I have worked with students to carve research topics out of the GRIP's scope of work. At the very minimum, this working together has helped ensure that students could have a territory for analysis and first authorship. In general, it has allowed me to work with students in projects where we are mutually invested in the substance of the work rather than only on the students' development. While students used data that had been or was being collected in response to broad research questions that I had formulated, the extent of their usage of such data corpuses has always summoned their agency in refining a question around a subfocus or creating a focus that could be examined using available data, and in any of those cases, their work has included original contributions to the framing of the specific problem or the research methodology. The continuous development of new research ventures has permitted my graduate students to participate in all phases of research (including instrument development and data collection) even if their own research project may have given them experience and relative autonomy only in analysis and writing. Overall I am proud of having been able to make that system work for 12 years and counting. I am especially proud of the five dissertations that, as of November 2013, have come out of the GRIP (Aaron 2011; Chen 2012; González 2009; Hamlin 2007; Weiss 2009), and the many opportunities for co-authorship and first authorship my students have had as they worked in the GRIP. Additionally, papers that most of those students

---

[2] The main corpuses of data are three: a video library of geometry lessons, a video library of focus group conversations among teachers talking about geometry lessons, and an extensive set of log data collected through our online platform, LessonSketch (www.lessonsketch.org).

wrote to transition to candidacy used data and ideas worked out in the GRIP and became published articles (Aaron and Herbst 2012; Chen and Herbst 2013; González and Herbst 2009; Weiss et al. 2009). I am only regretful that the numbers have not been higher.[3]

Jeremy's comments do give me a bit of pause as I think of my dissertation advising style, however. On the one hand, sustaining the kind of dissertation advising I have done has not been easy. My scholarly self-interest has been sufficiently engaged to the point that I have spent much time and energy with my students' dissertations, and for the most part, this has felt like working on ideas that I also wanted to push forward. But this work has often included intensity and frustration on both sides, some of which might be attributed to the kind of dissertation advising enacted. More concerning, however, is the number of students that I may have missed advising perhaps because of a lack of a clear interest match when assessing applications or too narrow expectations when recruiting admitted doctoral students. I have not made great efforts to recruit potential students whose interests were far from mine, and instead, I have described my style to potential recruits in rather explicit terms. I realize that that has been a choice, and one that was instrumental to maintaining my focus, but as I grow up in the field I wonder whether that choice is a liability of sorts, one that may jeopardize us as a field. I am not sure that it is, but I am happy that the interview with Jeremy provides some valuable fodder for such questioning.

My more than 10 years of reviewing applications to our Ph.D. program provide anecdotal evidence that is consistent with Jeremy's point about the need to focus the dissertation on individual development: Most applicants to doctoral programs in mathematics education are not ready to commit to a line of research. In fact, they have hardly read any research and tend to apply to programs with little regard to the specific work that scholars who teach in the program do. In other fields, it is common for graduate students to find places to study based on resident expertise on their area of choice. As Jeremy says, in mathematics education, this is seldom the case. Even international students, who may be applying to a program because of a person they want to work with, may not know much about the specific work that person does. This was, by the way, my case when I applied to Georgia—I did want to work with Jeremy and knew some of his work on problem solving, but I was not very interested in that topic and did not write about that topic when I wrote the personal statement in my application for admission to the University of Georgia. I did have specific interests that I had already been working on: I was then interested in the way the expansions of the number concept are taught and, particularly, on the ways textbooks represent the expansion of number systems up to the real numbers. Jeremy's student-centered style of advising allowed me to finish off that work as a master's thesis (Herbst 1995; see also Herbst 1997, 1998).

Most students that apply to doctoral programs have not really read enough to know what they are interested in. They come up with something that piques their

---

[3] Per my count, between 2001 and 2012, we've had 25 Education Ph.D. students graduate with a dissertation written in mathematics education (including mathematics teacher education), who were advised by 7 different faculty members.

interest, but usually don't feel strongly about it and hence are not prepared to make a commitment to it or to anything other than it. That situation is not ideal when we compare ourselves with other fields, but it is our reality: The pool of applicants to doctoral programs in mathematics education is not very large, but it is from that pool of applicants that we have to draw those who will continue our work as a field. It seems that we need to worry about recruiting and nurturing the best people regardless of whether they know what each of us is working on.

While other fields such as educational psychology can count on applicants being more informed of the research, deliberate in their application strategy and career orientation, and just more numerous, mathematics education does not seem to have reached a critical mass of aspirants to doctorates to be able to afford a laboratory approach in dissertation advising as the norm. Instead, it seems that, regardless of what kind of work they do in other aspects of their programs, we may still need to use the dissertation as an opportunity for individual development, not only of capacity for problem framing, field knowledge, and writing skills, but also of methodological skill, theoretical knowledge, and intellectual dispositions. A mixed style of dissertation advising may be helpful to the field as a strategy of human resource development. A laboratory approach may fit well those students who are interested and able to engage in the specific work that a particular group or scholar is doing at the time and reap the benefits of such close colleagueship in the form of more publications and access to resources. But with other students who aren't ready to make such a commitment, there may be ways in which a student-centered approach to dissertation advising could also be good for the mathematics education field. As fields like psychometrics, education policy, and economics have made inroads into mathematics education research, it seems really important for us to give serious consideration not only to the substantial progress of our individual areas of inquiry but also to the growth of the research capacity of native members of the field. We need to continue to be able to control how we conceptualize our object of study in mathematics education research, and this requires reproducing our capacity for methodological sophistication so as to be able to stay competitive in our quest for mathematics education research funding. It may be possible that those students who don't quite know what they are interested in could explore a variety of substantive topics while focusing instead on learning and using particular research designs, methods, and techniques that we, as a field, should be interested in keeping abreast of and having available. In that way, we might be able to reach a compromise between programmatic focus on the needs of the field and attention to the substantive interests of the individual student.

Conclusion: Whither Dissertation Advising in Mathematics Education?

A consideration of Jeremy Kilpatrick's advising style is of value to all of us. His voice persists in my head, saying that the matter of pride, the legacy of his work, is the student rather than the dissertation. We are at an important transition point in the field of mathematics education. The emergence of qualitative research methods some twenty plus years ago enabled the field to devote substantial energy to theory development. This work has not necessarily ended, but, in the meantime, some theoretical work has matured to the point of demanding increased methodological sophistication to be dedicated to construct measurement and theory testing. In parallel, technological advances have made possible the creation of large corpora of rich data, and research methods in the social sciences have become more adept at handling the phenomena in the teaching and learning of mathematics that we care about. But our human resource development approaches have not kept abreast of the range of designs, methods, and techniques available in the social sciences. We may be now in a situation of needing massive retraining to address our objects of study in the best ways. Clearly, a laboratory approach for dissertation advising can develop skill in some methods, but it may be too narrow-minded to enable our field to maintain theoretical and methodological control over our objects of study. The scholars that come out of our doctoral programs need to be able to envision new and better work or to bring sophisticated methodological knowledge to support existing work.

I take Jeremy's dissertation advising style, so open to the focus that the student may want to follow and so unabashedly loyal to the mathematics education field in his way of responding to the student's interest, as indicating a possible path ahead. Indeed, Jeremy has dedicated quite a bit of his writing to examine the history and condition of our field and to provide suggestions for its improvement (see, e.g., Kilpatrick 1992; Silver and Kilpatrick 1994). I too think that our dissertations need to serve the field—the simple fact that public institutions spend so much money supporting graduate students[4] is for me a strong reason to believe that an academic doctorate should not be seen as an individual credential for a better job but as a public investment in our capacity to generate knowledge. I don't know whether Jeremy agrees with that but what I think he does say is consistent with that position. I think Jeremy is really saying that doctoral education is about the human resources of the field at large rather than about specific research agendas: Doctoral education in mathematics education is about bringing the best minds we can to the field to maintain and hopefully increase the sophistication with which we understand mathematics teaching and learning. Could our best legacy be one in which we give in to whatever substantive focus the individual student wants to pursue but emphasize the theoretical sophistication and methodological skill development that a new generation of mathematics education scholars will need to have, in order to grow our field out of our present limitations and maintain ownership of our subject matter?

Acknowledgments I acknowledge the support of Elizabeth B. Moje, Associate Dean for Research at the University of Michigan's School of Education, who supported the transcription of the interview included here. I also acknowledge comments to an earlier version by Jeremy Kilpatrick, Wendy Aaron, Mike Bastedo, Kristen Bieda, Vilma Mesa, and Patrick Thompson.

---

[4] I calculate the cost of a doctoral student at the University of Michigan to be about $375,000 in five years counting tuition, year-round stipend, and benefits.

References

Aaron, W. (2011). *The position of the student in geometry instruction: A study from three perspectives*. Unpublished doctoral dissertation. University of Michigan, Ann Arbor.

Aaron, W., & Herbst, P. (2012). Instructional identities of geometry students. *Journal of Mathematical Behavior, 31*, 382–400.

Appleby, J. (2006). Historians, the historical forces they have fostered, and the doctorate in history. In C. Golde, G. Walker, & Associates (Eds.), *Envisioning the future of doctoral education: Preparing stewards of the discipline* (pp. 311–326). San Francisco: Jossey Bass.

Bass, H. (2006). Developing scholars and professionals: The case of mathematics. In C. Golde, G. Walker, & Associates (Eds.), *Envisioning the future of doctoral education: Preparing stewards of the discipline* (pp. 101–119). San Francisco: Jossey Bass.

Boud, D., & Lee, A. (2009). *Changing practices in doctoral education*. New York, NY: Taylor & Francis.

Breslow, R. (2006). Developing breadth and depth of knowledge: The doctorate in chemistry. In C. Golde, G. Walker, & Associates (Eds.), *Envisioning the future of doctoral education: Preparing stewards of the discipline* (pp. 167–186). San Francisco: Jossey Bass.

Chen, C. L. (2012). *Learning to teach from anticipating lessons through comics based approximations of practice*. Unpublished doctoral dissertation. University of Michigan, Ann Arbor.

Chen, C. L., & Herbst, P. (2013). The interplay among gestures, discourse and diagrams in students' geometrical reasoning. *Educational Studies in Mathematics, 83*(2), 285–307.

Golde, C., Walker, G., & Associates. (2006). *Envisioning the future of doctoral education: Preparing stewards of the discipline*. San Francisco: Jossey Bass.

González, G. (2009). *Mathematical tasks and the collective memory: How do teachers manage students' prior knowledge when teaching geometry with problems?* Unpublished doctoral dissertation. University of Michigan, Ann Arbor.

González, G., & Herbst, P. (2009). Students' conceptions of congruency through the use of dynamic geometry software. *International Journal of Computers for Mathematical Learning, 14*(2), 153–182.

Hamlin, M. L. (2007). *Lessons in educational equity: Opportunities for learning in an informal geometry class*. Unpublished doctoral dissertation. University of Michigan, Ann Arbor.

Herbst, P. (1995). *The construction of the real number system in textbooks: A contribution to the analysis of discursive practices in mathematics*. Unpublished Master's Thesis. University of Georgia, Athens.

Herbst, P. (1997). The number-line metaphor in the discourse of a textbook series. *For the Learning of Mathematics, 17*(3), 36–45.

Herbst, P. (1998). Metaphor and mathematical discourse. In J. F. Quesada (Ed.), *Logic, semiotic, social, and computational perspectives on mathematical languages* (pp. 43–63). Seville: SAEM Thales.

Jones, L. (2009). Converging paradigms for doctoral training in the sciences and humanities. In D. Boud & A. Lee (Eds.), *Changing practices in doctoral education* (pp. 29–41). New York, NY: Taylor & Francis.

Kilpatrick, J. (1992). A history of research in mathematics education. In D. Grouws (Ed.), *Handbook of research on mathematics teaching and learning* (pp. 3–38). New York: Macmillan.

Lester, F., & Carpenter, T. (2001). The research preparation of doctoral students in mathematics education. In R. Reys & J. Kilpatrick (Eds.), *One field, many paths: U. S. doctoral programs in mathematics education (CBMS Issues in Mathematics Education* (Vol. 9, pp. 63–66). Providence, RI: Mathematical Association of America.

Reys, R., & Kilpatrick, J. (2001). *One field, many paths: U. S. doctoral programs in mathematics education (CBMS Issues in Mathematics Education, Vol. 9)*. Providence, RI: Mathematical Association of America.

Reys, R., Glasgow, R., Teuscher, D., & Nevels, N. (2007). Doctoral programs in mathematics education in the United States: 2007 status report. *Notices of the American Mathematical Society, 54*(10), 1283–1293.

Sekiguchi, Y. (1991). *An investigation on proofs and refutations in the mathematics classroom.* Unpublished doctoral dissertation. University of Georgia, Athens.

Silver, E. A., & Kilpatrick, J. (1994). E pluribus unum: Challenges of diversity in the future of mathematics education research. *Journal for Research in Mathematics Education, 25*(6), 734–754.

Weiss, M. K. (2009). *Mathematical sense, mathematical sensibility: The role of the secondary geometry course in teaching students to be like mathematicians.* Unpublished doctoral dissertation. University of Michigan, Ann Arbor.

Weiss, M., Herbst, P., & Chen, C. (2009). Teachers' perspectives on "authentic mathematics" and the two-column proof form. *Educational Studies in Mathematics, 70*(3), 275–293.

Chapter 3
Collaboration and Friendship with Jeremy Kilpatrick: Two Sides of Success and Challenge

Christine Keitel-Kreidt

Abstract I relate the story of my first personal encounter with Jeremy Kilpatrick and trace how it led to several wonderful collaborations, beginning with a book co-written with Jeremy and with Geoffrey Howson. The story reveals both Jeremy's intellectual curiosity and his supportive colleagueship.

Keywords Mathematics • Curriculum • Collaboration

When meeting American colleagues, many of them praise the intellectual support and encouragement they have received from Jeremy Kilpatrick. And they always reminded me that I am also such a colleague, who has been challenged and supported by Jeremy Kilpatrick to produce research analyses and engage in theoretical analyses and problem development.

It was in 1975, the newly established "German Institute for Didactics of Mathematics" at the also newly founded University of Bielefeld started a first "world conference on mathematics education (didactics)" that called in about 75 especially invited and distinguished colleagues from all over the world, mainly from the USA, the UK, France, and few other countries. At the occasion of the Bielefeld Conference on Math Education, I met personally Jeremy Kilpatrick – so far I only knew him by his publications. But I learnt that he was very much curious about other peoples' views on math education and all kinds of different theoretical approaches and analyses.

Working at the Max Planck Institute for Educational Research, I had been asked by the organizers to deliver a paper on my analysis of the most recent curriculum reforms and describe and analyze in particular the huge variety of approaches and material support developed mainly in the USA and Great Britain and some other countries. I had analyzed and comprised several conceptualizations as differently designed approaches that I had identified and aimed at a substantial overview of these developments including a substantial analysis and accompanying concrete examples of possible "chances and failures".

C. Keitel-Kreidt (✉)
Fachbereich Erziehungswissenschaft und Psychologie, Freie Universität Berlin,
Berlin, Germany
e-mail: keitel@zedat.fu-berlin.de

© Springer International Publishing Switzerland 2015
E. Silver, C. Keitel-Kreidt (eds.), *Pursuing Excellence in Mathematics
Education*, Mathematics Education Library, DOI 10.1007/978-3-319-11952-6_3

After my presentation, the first reaction from one English colleague, one who belonged to the working group organizers of ICME 1976 in Karlsruhe, was to invite me into his presentation – and discussion group to present there – at that time it was against the German "academic rules" that a non-PHD-holder should be allowed as a presenter, but my foreign colleagues succeeded to keep me as a presenter. The surprising experience I made that many American and British colleagues approached me and wanted to get more about "my interesting presentation," actually I only had just designed it as the theoretical frame as part of my work in progress for the doctoral thesis.

The final result of my presentation and the very successful public debate with many colleagues was an unexpected offer by two especially distinguished math educators who asked me to collaborate with them on a book that could integrate and relate my conceptualizations of the various curriculum reform approaches and collect and analyze the new teaching and learning material as well. I was asked to enlarge the newly developed conceptions of teaching and learning to design an international volume.

This was the birth of "Howson, Keitel, Kilpatrick – Curriculum Development in Mathematics." The book was published by Cambridge University Press in 1981, and until today, it has already got several reeditions, the last from 2010. Further on, when I visited Japan and China, with great pride, Japanese and Chinese colleagues presented me the translation of "Curriculum Development in Mathematics" into Japanese and Chinese, as well as a Spanish version, and many others congratulated to my successful collaboration with two distinguished and famous colleagues. I was an "unknown" so far, but colleagues did tell me how much they like "our" book.

The book would not have come into being if Jeremy and Geoffrey had not persuaded me. We planned and realized several meetings in Southampton to work on the book; I was pregnant, so my husband Fritz joined us to support me.

After long and detailed discussions, we decided to design it also as a working book: working on ideas and problems when reading and creating or developing new ideas. This structure of the book actually followed closely my ideas, which I had developed together with a timeline of the creation of the various conceptualizations.

One very important idea of Jeremy's was the invention and inclusion of "problems" – closely related to the different approaches and conceptualizations – to work on when reading, either directly related to the various essays on curriculum development or to specific theoretical aspects and so on, whoever has seen or even used the book knows about its rather exceptional design as a mixture of reading book with problems to solve on the way through. Still today, I have great memories of this amazing collaboration between a complete newcomer and two world-class colleagues! But it was the best way to quickly learn and develop, to enlarge my intellectual capacity, and to adapt or appropriate new ways to learn and write – including as well to also improve my English!

Later on we have worked together in various working groups on conferences, we are successful in several common publications and also in the Working Group BACOMET (Basic Components of Mathematics Education for Teachers) that met for several years at various places, and until today, we have got quite a nice collection of common publications. I am very thankful to Jeremy Kilpatrick for all his challenges and support!

Part II
Mathematical Problem Solving
and Curriculum

Chapter 4
When Is a Problem? Contribution in Honour of Jeremy Kilpatrick

John Mason

Abstract Some of Jeremy Kilpatrick's contributions to mathematics education are used to contextualise the question of 'when is [something] a problem?' leading to different approaches to or value systems for engaging students in mathematical thinking. My own preferred approach is experiential and hence phenomenological, in which the reader is invited to compare their experience of working on some task exercises with my commentary. The product of such enquiry is possibly revised and honed task exercises which can be used to sensitise others to the same issues.

Keywords Problem • Problem solving • Activity • Tasks • Phenomenology

Introduction

My encounters with Jeremy Kilpatrick have consisted of a Fulbright exchange visit between our universities in the early 1980s and all too brief meetings at international conferences. Jeremy has been a focal point for our community both for maintaining contact with the historical roots of mathematics education research and for a vision of fruitful directions for the way forward. He has continued to provide the community with valuable overviews of mathematics education through edited collections of papers, through his plenary addresses to ICME in 1984 in Melbourne and in 2008 in Mexico, and in his recent paper providing historical reflection on the phenomenon known as the 'new math' (Kilpatrick 2012).

Jeremy's thesis was about problem solving, supervised by George Pólya. I recall coming across a copy in a library in the mid-1980s and being relieved to find that it was entirely consonant with my own experience as expressed in Mason et al. (1982–2010). However, Jeremy's approach to research is what I am tempted to describe as traditional in its extra-spective[1] focus, and this contrasts with my more

---

[1] Observing from the outside, that is, observing others.

This title is taken from Brookes (1976).

J. Mason (✉)
Department of Education, Open University and University of Oxford, London, UK
e-mail: john.mason@open.ac.uk

© Springer International Publishing Switzerland 2015
E. Silver, C. Keitel-Kreidt (eds.), *Pursuing Excellence in Mathematics Education*, Mathematics Education Library, DOI 10.1007/978-3-319-11952-6_4

intra-inter-spective[2] stance (Mason 2002). The thinking reported in this chapter reflects my own approach, which is experiential or, in more philosophical terms, phenomenological. I begin every enquiry by interrogating my own recent experience and looking for mathematical situations in which I can experience the phenomenon of interest freshly. I then present honed versions of those situations as task exercises to others with a view to checking whether they notice similar phenomena or can make similar distinctions. Negotiating the language of description of that experience usually leads to some technical terms that can be used for description and analysis of further instances of the phenomena in the future. Validity lies in whether readers find their own choices more richly informed in the future.

This approach was outlined in Mason (2002) as the Discipline of Noticing, because it seemed to me that the person who gains most from research is the researcher themselves, as I proposed (Mason 1998) in one of the many collections that Jeremy has edited. Research in mathematics education (and more generally in the social sciences and beyond) is based on researchers discerning ever-finer distinctions in what they observe, and to my way of thinking, this can best be grounded in and through personal experience. I have more than once chided Jeremy that in his overviews of the state of play in mathematics education, he consistently overlooks the phenomenologically experiential basis of all research, particularly in mathematics education, and, even more importantly, the contribution that can be made by adopting it as an explicit stance.

'Problem' as Problematic

One of the weaknesses of mathematics education as a domain of enquiry is that there is no agreed practice of negotiating meaning of technical terms (Mason 2011). For example, Keitel and Kilpatrick (2010, pp. 105–106) point out the increasingly frequent use of the term 'common sense' with extended uses and meanings based on variations such as 'everyday common sense', 'situated knowing', 'cognition in practice', 'everyday cognition', 'a priori understanding', 'intuition' and 'knowledge based on subjective experience'. More abstract but related terms include 'children's mathematics' and 'cultural perspectives' as well as a changing view of 'common sense about mathematics' historically. My aim is to convert 'knowledge based on subjective experience' into evidence-based knowledge that can be and is shared in the community and that informs future practice. Similarly, it is becoming a common awareness that children can solve problems that are supposed (by authors, teachers and curriculum developers) to be beyond them, especially in the younger years, before they have become enculturated into thinking that they have to be told how to do something in order to do it. Keitel and Kilpatrick (*op cit*. p. 106) use as their evidence a Dutch study (van den Heuvel-Panhuizen 1994) but the same applies to many students classified as 'low attaining' (see, e.g. Silver and Stein 1996; Boaler

---

[2] Observing oneself and negotiating these observations with the self observations of others.

1997 and Watson 2007). The concept of what constitutes a 'problem' is a prime example of a spectrum of meanings that cloud the effectiveness of research. This is taken up in the next section.

One consequence of failure to be precise about the meaning of technical terms is that researchers often use terms in a slightly special or idiosyncratic way, which makes it hard for the field to get beyond the proliferation of technical terms. Another consequence is that researchers make ever-finer distinctions in order to be seen to be contributing to the field, but increasingly often these distinctions are far finer than is helpful to the practitioner in the classroom. When research reports lose sight of the experience of students and classroom practices, they enter an abstract world of their own. Without a grounding in experience, there can be little impact where it really matters, which is students' experience.

Problems in the Curriculum

There is an English adage that 'what goes around, comes around', and nowhere is this more evident than in 'problem solving' as a focus of attention in mathematics education. It was a major focus when I began to be involved explicitly in the late 1970s and early 1980s, sometimes coupled with 'modelling' and with 'real problem solving' (Open University 1980; Mellin-Olsen 1987). It was largely displaced by assessment and then returned again, only to subside and then re-emerge recently (see, e.g. Lesh and Fennewald 2010, or Badger et al. 2012). In his inestimable manner, Kilpatrick (2012) gives a potted but comprehensive survey of the multiple reforms commonly subsumed under the banner of 'new math', pointing out the novelty in the 1960s of the very notion that a mathematics curriculum could be capable of reform, much less actually requiring it. Previously, the mathematics curriculum seemed as entirely 'natural' as it was 'necessary'.

> Before t;he new math era, no one thought of school mathematics as something to be reformed or updated; it simply was what it was. (Kilpatrick 2012, p. 564)

> However, apart from Dieudonné proposing that

> The student should already be fairly well trained in the use of logical deduction and have some idea of the axiomatic method. (Quoted in Kilpatrick, p. 563)

no explicit mention was made of 'problem solving'. This is an example of 'when is a problem' at the curricular level: until the sputnik fright the curriculum itself was not problematic, and until the curriculum became problematic, neither the role of problem solving nor the meaning of 'problem' could be questioned. In his account of the positive and negative features of the then current research in problem solving, Jeremy provided indications of possible future developments, most of which have materialised, but without, in my view, repairing any of the weaknesses he discerned (Kilpatrick 1978).

Alan Schoenfeld undertook a review of 'problem solving' somewhat akin to Kilpatrick's review of curricular reform (see also Silver 1985). In the late 1970s and early 1980s, there were calls in the USA for emphasising 'problem solving' (Schoenfeld 1992, p. 8) and it became a theme for curricular reform in the 1980s. It moved from an 'unusual aspect of mathematics at ICME 1978 to one of the seven principal themes at ICME 1982' (Schoenfeld 1992, p. 8). In common with other authors, Kilpatrick and Schoenfeld drew attention to the problematic nature of the word 'problem' itself, pointing out that what people mean by 'problem' varies. At one end of a spectrum are traditional textbook questions, found in ancient Babylonian, Egyptian and Chinese texts and continued to the present day, which usually demand a specific answer. At the other end of that spectrum are explicitly exploratory or investigative nonroutine tasks which now go under the heading of 'rich' or 'open-ended' tasks. Extreme versions of these are novel and unsolved problems which lie at the heart and core of what professional mathematicians do, for example, as expressed by Paul Halmos:

> The mathematician's main reason for existence is to solve problems, and ... therefore, what mathematics really consists of is problems and solutions. (Halmos 1980, p. 519) (Quoted in Schoenfeld 1992, p. 15).

The place of 'problem solving' within the curriculum depends not only on the meaning of 'problem' but also the stance taken towards mathematics itself. From Pólya's perspective,

> If the learning of mathematics has anything to do with the discovery of mathematics, the student must be given some opportunity to do problems in which he first guesses and then proves some mathematical fact on an appropriate level. (Quoted in Schoenfeld 1992, p. 17)

Brookes (1966) directly addressed the question of the place of problems in the curriculum, reflecting views held by many members of the Association of Teachers of Mathematics founded by Caleb Gattegno, which had a significant influence on people such as John Holt and Bob Davis. Students can truly be said to be learning mathematics when they adopt a mathematical stance towards the world:

> Seeing the world from a mathematical point of view, as problematic, but in a particular 'mathematical' way. (Schoenfeld, pp. 22, 31)

This involves problem posing as well as problem solving. Indeed the two must go hand in hand if either is to prevail (Brown and Walter 1983; Silver 1994).

In Boaler's original research there was evidence that the students engaged in project work reported elements of this sort of stance: the world began to become mathematically problematic for them. Cuoco et al. (1996) added considerably more detail to the behaviour of mathematically minded people in their description of mathematical *habits of mind*.

Experiential Approach

I take the stance that mathematics is about posing and solving problems. For me something is a problem only when someone experiences something *as* a problem. What interests me here is the transition between different psychological states, at least one of which could be called 'having a problem', or 'being in a problematic state'. By 'psychological state' I mean a particular combination of cognition, affect, behaviour and attention, often associated with a particular 'self' but sometimes involving a struggle between different 'selves'. The stimuli that follow in this section are task exercises (Mason 2002). They are designed to provoke activity; if carried out, that activity will generate experience (if not, there is no point in reading further!); within that experience it may be possible to notice transitions from 'task' to 'activity' to 'problematicity'. My comments are mainly, but not exclusively, directed towards features of states of 'being problematic' in the sense of Heidegger (1927), who used a crossed-out version (being) to indicate that 'being' is not some 'thing', some aspect which can be detected and measured. Rather, it is a portmanteau for 'being in the situation', sensitised and responsive to what emerges. The more common term 'identity' interpreted as 'who and how one is' partly signals the same idea but with overtones of self-awareness and ego, whereas 'being' has a deeper philosophical foundation in ancient psychology.

Circle Dissection

Dissect the first circle shown here into four congruent pieces.

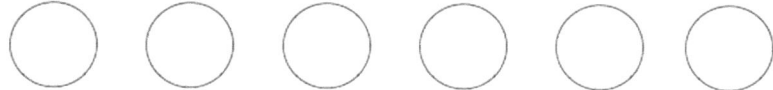

Now do it in a different way.
Now do it in a yet different way.
How do you know whether or not there is another different way?
What do you mean by 'way' of doing the dissection?

Comment

The first request is entirely unproblematic. Almost everyone immediately draws a cross using straight lines. When asked to do it again differently, some people rotate the cross but most people trace a wiggly route from the centre to the circumference and then repeat it three more times at angles of 90°. Some people already ask 'what do you mean by different?'. For these people the focus of attention has shifted slightly to 'what does the teacher want?' or to 'what are the rules, the constraints under which I am operating?'. The third request raises the 'what counts as different?' question even more sharply as people recognise that they now see all of their solutions as being of the same type: they are stressing what is the same and ignoring or backgrounding what is different. Some people are immediately aware of a generality (they can take any route from the centre to the circumference and repeat it four times, though they may as yet be unaware that there are constraints on the route!). The fourth question raises the issue of justifying conjectures by using mathematical reasoning, with an implication that there is in essence only one way to dissect a circle into four congruent pieces, where 'way' has to do with clarifying what is meant by 'route'. The fifth question focuses attention on this clarification. Now the issue is meaning of 'route' not simply the construction of a single route.

Different people become engaged, experience the situation as problematic, at different points in the unfolding. If the last two questions are taken up, then contact is made with the difficulty in providing a completely convincing justification, as it is hard to pin down exactly what to base your reasoning on other than that it seems essential that the centre be on all four pieces.

The mathematically inclined thinker explores what Marton and Booth (1997) call 'dimensions of variation' which Watson and Mason (2005) extended to 'dimensions of possible variation': changing the number of pieces and changing the circle to other shapes to try to see what is going on. The interested reader might like to try dissecting a circle into 12 congruent pieces, which can be done in several different ways. The question of whether a circle can ever be dissected into a finite number of congruent pieces with the centre in the interior of one of the pieces remains unresolved so far as I am aware. At different times this has been a 'problem' for me in that I have worked hard at trying to resolve it, but it is now in the category of 'as yet unresolved but probably too difficult'.

The point of this example of a potentially problematic situation is to observe *when* it becomes a problem, namely, the confluence of a sense of gap or chasm with no immediately obvious route to a resolution and a desire to find such a resolution. Something is problematic when all aspects of the psyche are engaged: cognition (a sense of gap or chasm), affect (desire to bridge that gap or to 'fill it in'), behaviour (no immediate route available) and attention focused on something specific, not just 'wondering'.

This is where the notion of *heuristics* comes in, a term introduced by Pappus and resuscitated by Pólya (1945, 1962). When something is indeed a 'problem', you make recourse to strategies, to the use of natural powers, that is to ways of thinking mathematically, such as outlined in Mason, Burton and Stacey (1982/2010) as a simplified version of Pólya.

One Sum

> I have two numbers that sum to one. I am going to square the smaller and add the larger; I am also going to square the larger and add the smaller. Which of my two answers will be bigger?

This particular version invites a conjecture by invoking a generality, immediately placing the reader in a state of not knowing the numbers. It might trigger the use of specialising to see what sorts of things happen and to get an inkling of which result will be the larger. Surprise depends on people being aware that there are lots of choices.

An alternative setting is to ask people to write down two numbers that sum to one, to ask them to perform the computations, and then to announce that the difference between their two answers is zero. In this version, surprise depends on the universality, the fact that the difference is independent of the choices made by individuals. This in turn depends on people choosing different numbers. Experience suggests that many people choose one and zero, or a half and a half, and so there is no surprise.

One way to combine the effectiveness of both settings is to begin by asking people to write down a pair of numbers that sum to one, then another different pair, then another different pair. Getting people to reveal their third pair is likely to bring to everyone's attention that there are many different possibilities. The strategy of inviting students to construct their own mathematical objects and to ask for 'another and another' turns out to be very effective in extending the accessible example space within which students are thinking (Watson and Mason 2005). Once the fact of multiple possibilities is established, moving on to the calculations produces an extra frisson of surprise when the results come out to be the same.

Choosing among the slight variations in presentation or setting so as to maximise the possibility that people will experience a gap, so that they will experience something as problematic, is part of the art of teaching. There is no universal rubric that will be effective every time. Rather, choices can be informed by being sensitive to student experience and particularly to the likely example spaces that will come to mind in response to the words or diagrams used.

One of the features of this task that appeals to me and that works well with students who either have not yet met algebra or who have resisted or rejected algebra as meaningful to them is that they can be induced into thinking algebraically without realising it. Using a cloud or speech bubble to stand for what some specific person is thinking is culturally familiar from cartoons. Often it helps to refer to someone who is not present so that they cannot be asked for their number. The cloud can then be used in the way that Mary Boole (Tahta 1972) advocated, as an acknowledgement of ignorance as to the number, but a way to present the as-yet-unknown to yourself so that you can express what you do in fact know. I have had students who have rejected algebra and become enthused about it as a result of working on this task. Before doing computations with the cloud, however, I usually invite them to interpret the computations diagrammatically as shown below.

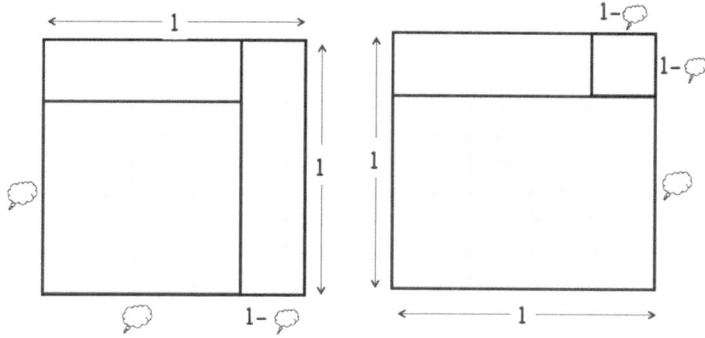

There are of course multiple ways to initiate interpretation of the diagrams, for example, building them up step by step or showing one and then the other, asking students each time to 'say what they see' and, when seeing both together, 'What is the same and what is different about the two diagrams?' (Mason et al. 2005). Again the basis for choice is sensitivity to students' current states within the usual classroom rubric, what they are attending to and in what way. For example, when asking what the second number would be if the first was 'cloud', I have several times been offered 'lightning bolt', evidence that the person was generalising well, but not attending to or making use of known relationships. Such a response can be interpreted as indicating engagement with the situation and a desire to contribute, although it could also be an attempt to be superficially facetious. Voice tones, posture and gesture can give clues as to which interpretation is appropriate, as can subsequent engagement.

The diagrams appear to deal only with a pair of positive numbers that sum to one. Various authors have sought ways to include negative lengths and negative areas into such diagrams (see, e.g. Sawyer 1959). One way, consistent with cross products of vectors, is to compute areas by always multiplying lengths in an anticlockwise direction (so clockwise includes a negative sign). Even when reinterpreting diagrams so as to permit negative lengths by keeping track of orientation is out of reach, the possibility remains to check the result arithmetically in specific instances and to see that the 'algebra' deals with both positive and negative values. This could contribute to accepting algebra as more powerful than diagrams for working with as-yet-unknowns. It could also create a wonder as to how diagrams could be extended to cope with negative lengths.

For previously algebra-rejecting students, this task may be sufficient for 1 day, but there are several variations that can be returned to and developed over time or presented immediately to more algebraically sympathetic students. For example, suppose the two numbers add up to 10 instead of 1. How must the original calculations be changed so as to preserve the equality? This provides experience of what Lakatos (1976) refers to as 'preserving the phenomenon'. Meanings of definitions are extended so as to preserve the 'theorem'.

In a different direction, taking the number of numbers as a dimension of possible variation, what about three numbers that sum to one. It turns out that moving to three or more numbers summing to one provides an opportunity to work with interchanging summation signs. If the algebraic expression is mysterious, then staircase-like diagrams (see below) can provide a source, something to interpret.

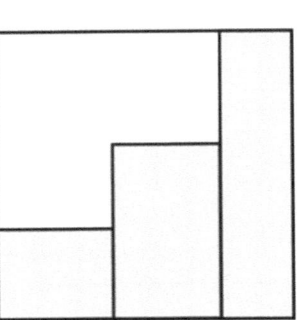 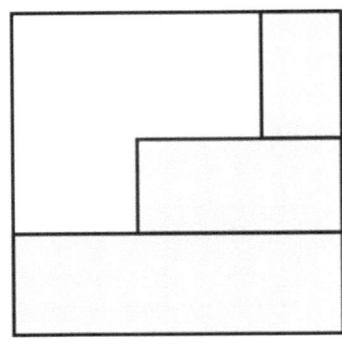

Meanwhile students have been exposed to an important mathematical principle that expressing the same thing in two different ways can produce unexpected and useful results.

In a different direction again, moving into three dimensions, consider the two diagrams below:

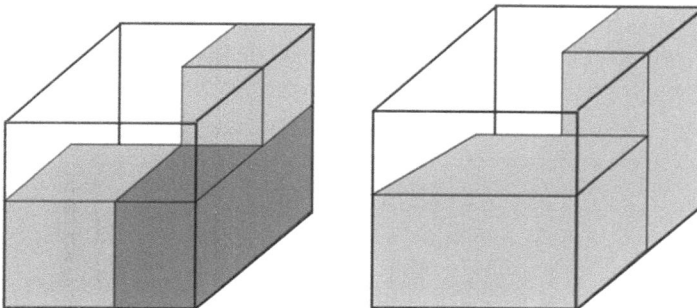

The outer cubes have unit lengths in each direction. Again take two numbers that sum to one and calculate firstly the sum of their cubes and their product and secondly the sum of their squares. Alternatively, interpret the diagrams as a statement of the equality of two expressions. Then look for dimensions of possible variation to alter. For example, suppose you have two pairs of numbers that sum to one, or even three pairs of numbers that sum to one, so that what looks like a cube is actually a cuboid. You could also try changing the sides of the outside cube or cuboid.

That might lead you back to the two-dimensional situation using two pairs of numbers that sum to one. Take the product of the first of each pair and add the second of the first pair. Compare that with the product of the second of the two pairs added to the first of the second pair. Surely it is easier to use icons like clouds or, even more simply, symbols of some sort, in order to present these as-yet-unknown numbers and calculations involving them. Returning, finally, to three or more dimensions, what is the three-dimensional analogue of the two-dimensional staircase diagrams?

Each of these variations was attractive to me at the time of posing precisely because I posed them myself, admittedly often with the thought that I might then pose them to others. The contrast between 'problems' posed for oneself with 'tasks' set by a teacher or text is psychologically significant. People who are predisposed to take on intellectual challenges may not experience this contrast as starkly as people not yet so disposed. By 'being mathematical with and in front of students', that is, by displaying a problem-posing stance, by parking some problems with conjectures for taking up at a later time (perhaps never) and by using increasingly indirect prompts to provoke students into adopting a problem-posing stance, more students will experience 'a problem' than if students only ever suffer (in the original sense of the word) tasks administered by a teacher from a prepared stock.

Marbles

Here are some stimuli offered in a sequence so that the reader can try to detect when or where there is a shift from the non-problematic to the problematic. It may be that none are problematic in the sense that someone competent with algebra will know that (or at least assume that) they can resolve all of them. Nevertheless there will be a point at which there is a shift from the 'it's immediately obvious' to 'it needs some effort'. It will then be possible to notice an experience of shift of disposition as to whether to take on the challenge or not.

> If Jeremy were to give John three of his marbles, they would have the same number of marbles. How many more marbles does Jeremy have than John?
>
> If Jeremy were to give John one of his marbles, Jeremy would then have one more than twice as many marbles as John then has. However, if, instead, John were to give Jeremy one of his marbles, he would have one more than a third as many marbles as Jeremy then has. How many marbles do they each have now?
>
> If Jeremy were to give John seven of his marbles and then John were to give Ed three of his, they would all have the same number of marbles. What is the relationship between the number of marbles each currently has?
>
> If Jeremy were to give John seven of his marbles and then John were to give Ed three of his, Jeremy would have twice as many marbles as John and three times as many marbles as Ed. What is the relationship between the number of marbles each currently has? What is the smallest number of marbles that Jeremy could have?
>
> If Jeremy were to give John seven of his marbles, John were to give Ed three of his marbles, and Ed were to give Jeremy one of his marbles, they would all have the same number of marbles. What if Jeremy were then to have twice as many marbles as John and three times as many as Ed?
>
> If Jeremy were to give seven marbles to John, John would have as many marbles as he would have had if Ed had given him eleven marbles more than the number Ed would need to give Jeremy so that Ed had one-third as many marbles as Jeremy would have then had. But then Ed would have had half as many marbles as John. What relationships are there between the numbers of marbles each currently has?

Comment

The first stimulus has an analogue with voting changes: the effect is doubled. Notice the potential generality achieved by changing the number of marbles Jeremy might give to John. All of these stimuli have analogues with the collection of 'age' problems ubiquitous in eighteenth and nineteenth century arithmetic and algebra books. An alternatively probe would be to ask what can be said about the relationship between the number of Jeremy's marbles and the number of John's marbles, as in the later stimuli. This would help emphasise that arithmetic is about relationships between numbers, stressing that it is the number of marbles of interest, not their colour or size.

Set in terms of marbles, it is quite likely that students who want something to manipulate could use counters or even marbles, while cast in terms of age it might not be quite so obvious to use counters or positions on a number line to present oneself with confidently manipulable objects with which to investigate the situation (to model). Students can be invited to draw diagrams to capture (articulate) their thinking, and this can develop into a more formal verbal and symbolic format. The threefold framework *manipulating – getting-a-sense-of – articulating* (MGA) was developed (Open University 1982) as a reminder to teachers that specialising by turning to a confidence-inspiring example or instance or (re)presentation can provide something to do when you get stuck. However, the purpose of specialising is to re-generalise for yourself by getting a sense of underlying structural relationships. Thus, manipulating objects is not an end in itself but a means for contacting mathematical structure which you then try to express in various ways, to articulate in actions, pictures, diagrams, words and symbols. As the articulation becomes succinct yet meaningful, it becomes available in the future to be used as increasingly confidently manipulable entities for further thinking. MGA then becomes a tool for informing lesson preparation as well as a reminder of something to do when stuck on a problem.

The various versions of marble problems are intended as indicators of opportunities to play, to recognise and make changes in dimensions of possible variation and to generalise, thereby running into significant 'problems' in number theory concerning common divisors and articulating constraints on parameters so that a resolution is actually possible.

Reflections

I follow Bennett (1993) in seeing activity as structured around two axes, *motivation* and *operation*, taking place within a world of attention. Motivation in an activity has to do with the perceived gap between goals or aims and current state.

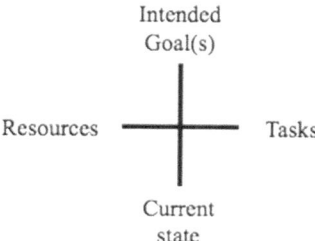

This is consistent with a Vygotskian perception of activity but with the addition of a tension or gap between current state and goal, which plays the role of disturbance identified by Heidegger (1962) and many others (e.g. Festinger 1957; Piaget 1971) as what activates learning (leading to assimilation and accommodation).

The second axis consists of the resources available (both those brought by the students and those provided by the environment) and the tasks provided. If the resources available are inadequate for the gap between current state and goal, if the tasks do not actually provide sufficient stimulus to reach the goal or if the gap is too slight for the resources to hand, then the activity will be ineffective.

Resources include student propensities and dispositions and learner access to their natural powers such as stressing and ignoring, imagining and expressing etc. Where student powers are usurped by textbooks or modes of interaction with the teacher, students soon learn to park their powers at the door as not being required and so become dependent on the teacher to initiate mathematical actions.

Tasks are inherently multiple by nature: as conceived by the author, as intended by the teacher, as construed by the student(s), as enacted by the students and as recalled in retrospect by the student(s). In the language of affordances, constraints and attunements (Gibson (1979), affordances arise from the relationship between resources and tasks. The constraints are usually imposed from the tasks, for as is well known, creativity only takes place when there are constraints. Student and teacher attunement contribute to the motivational and the operational axes.

The purpose of activity is to generate experience, but for most people most of the time, turning that experience into learning requires withdrawing or drawing back from the activity and considering what actions were successful and what actions were unsuccessful. In his typically foresight-full article reviewing problem-solving research in the late 1970s, Kilpatrick (1978, p. 192) expressed this by supplementing Dewey's dictum: 'we learn by doing' with 'and by thinking about what we do'. Even the notion of 'thinking about' requires refinement before it is likely to be effective. As Jeremy's colleague Jim Wilson was fond of saying, 'Pólya's fourth stage ('looking back') is more honoured in the breach than in the execution' (personal communication).

Disposition is an important thread in the braid that comprises understanding (Kilpatrick et al. 2001), but it is notoriously difficult to adjust. Slogans such as 'real problem solving' and 'authentic mathematics' are associated with a stance that equates 'willingness to engage' with 'appreciation of future use' and with the derivative notion that in each lesson students should be explicitly told the 'learning objectives'. Students are seen as requiring both future utility and local purpose (Ainley and Pratt 2002) before they will engage with challenges. But the question from students 'why are we doing this?', like the same question asked by people at different times in their life outside of school, usually signals not a desire to know where something can be used or applied but rather is a statement that confidence is ebbing. As Dweck (2000) has shown so convincingly, an epistemological stance based on 'intelligence is genetic' can be altered to one based on 'intelligence is like a muscle', if the language of 'can't' and 'won't' are converted to 'didn't' and 'could try harder or differently'. The art of teaching is to gauge an appropriate level of challenge in an atmosphere in which students trust the teacher to set them appropriate challenges. If the teacher is not constantly treading the edge between accessible and inaccessible, between achievable and unachievable, then students are being trained to be dependent on the teacher and so are not truly learning.

Kilpatrick (2003, pp. 319–320) describes the perennial tension between the practical and the intellectual in terms of two tectonic plates forever bumping against each other, with the current curriculum at any time being an articulation of the interface. There have been, are and must continue to be numerous attempts in every generation to articulate the ways in which these 'plates' interact and what each offers to students, if mathematics education is going to improve the experience of students.

As living organisms, human beings, certainly as children, respond to challenge that is perceived as accessible and even achievable. Unfortunately the institution of school often contributes to the suppression of this natural response. Children become inured to being told exactly what to do to the point of dependency. When curricula mention or even stress 'problem solving' (e.g. as in Singapore), it is not simply to 'get answers to routine tasks' but rather to develop and refine students' natural powers and dispositions to engage so that they respond creatively and flexibly to changing conditions:

> It is a major goal of mathematics education to help the formation of active, autonomous, creative, and flexible individuals. One aspect of this is the ability to identify, pose, formulate, handle, and ultimately solve problems that one encounters on one's way. (Niss 2003, p. 275)

It is my experience that where there is trust in the teacher, students can be intrigued without requiring immediate utility and that they can and will engage with challenges without evident 'purpose' beyond that of challenge. Student trust in the teacher is a reflection of teacher trust in the student (and of institutional trust in the teacher). This mutuality is something that begins freshly each year and is amplified or diminished in each encounter.

So the question of 'when is a problem?' is not easy to answer. I hope that in working on at least some of the stimuli presented here, the reader has noticed different and shifting attitudes (both in the affective and in the mathematical sense of the term). Sometimes desire is felt strongly, an urge to reach a resolution; sometimes desire is overlaid with other concerns (to finish the chapter, to 'get to the point'); sometimes desire is muted, dominated by a sense that you 'see where he is going or what he is pointing to'; sometimes it is backgrounded, as in my problem with dissecting the circle. As Heidegger (1927/1949) observed, in common with many others (e.g. Festinger 1957; Piaget 1971), without disturbance to the equilibrium there is no change, no need to respond, no learning.

Care is needed however! Disturbing intentionally, as an explicit aim, can be dangerous; disturbing as a consequence of being informed mathematically, pedagogically and didactically can be effective. There is no need for the teacher to disturb intentionally, as long as their Heideggerian mathematical being is active in informing their actions so that they are being mathematical both with and in front of their students.

References

Ainley, J., & Pratt, D. (2002). Purpose and utility in pedagogic task design. In A. Cockburn & E. Nardi (Eds.), *Proceedings of the 26th annual conference of the International Group for the Psychology of Mathematics Education* (Vol. 2, pp. 17–24). Norwich: PME.

Badger, M., Sangwin, C., & Hawkes, T., with Burn, R., Mason, J., & Pope, S (2012). *Teaching problem-solving in undergraduate mathematics.* Coventry: National HE Stem Programme, Coventry University & MSOR. Obtainable from www.mathcentre.ac.uk/problemsolving

Bennett, J. (1993). *Elementary systematics: A tool for understanding wholes.* Santa Fe: Bennett Books.

Boaler, J. (1997). *Experiencing school mathematics: Teaching styles, sex and setting.* Buckingham: Open University Press.

Brookes, W. (Ed.). (1966). *The development of mathematical activity in children: The place of the problem in this development.* Derby: ATM, Nelson.

Brookes, B. (1976). Philosophy and action in education: When is a problem? *ATM Supplement, 19,* 11–13.

Brown, S., & Walter, M. (1983). *The art of problem posing.* Philadelphia: Franklin Press.

Cuoco, A., Goldenberg, P., & Mark, J. (1996). Habits of mind: An organizing principle for mathematics curricula. *Journal of Mathematical Behavior, 15,* 375–402.

Dweck, C. (2000). *Self-theories: Their role in motivation, personality and development.* Philadelphia: Psychology Press.

Festinger, L. (1957). *A theory of cognitive dissonance.* Stanford: Stanford University Press.

Gibson, J (1979). *The ecological approach to visual perception.* London: Houghton Mifflin.

Halmos, P. (1980). The heart of mathematics. *American Mathematical Monthly, 87*(7), 519–524.

Heidegger, M. (1927/1949). *Existence & being* (W. Brock, Trans.). London: Vision Press.

Heidegger, M. (1962). *Being and time* (J. Stambough, Trans.). New York: Harper & Row.

Keitel, C., & Kilpatrick, J. (2010). Mathematics education and common sense. In J. Kilpatrick, C. Hoyles, & O. Skovsmose (Eds.), *Meaning in mathematics education* (Mathematics education library, Vol. 37, pp. 105–128). New York: Springer.

Kilpatrick, J. (1978). Research on problem solving in mathematics. *School Science and Mathematics, 78,* 189–192.

Kilpatrick, J. (2003). Scientific solidarity today and tomorrow. In D. Coray, F. Furinghetti, H. Gispert, B. Hodgson, & G. Schubring (Eds.), *One hundred years of l'enseignement mathématique: Moments of mathematics education in the twentieth century.* Proceedings of the EM–ICMI symposium (pp. 317–330). Geneva: L'Enseignement Mathématique.

Kilpatrick, J. (2012). The new math: An international phenomenon. *ZDM, 44,* 563–571.

Kilpatrick, J., Swafford, J., & Findell, B. (Eds.). (2001). *Adding it up: Helping children learn mathematics* (Mathematics Learning Study Committee). Washington, DC: National Academy Press.

Lakatos, I. (1976). *Proofs and refutations: The logic of mathematical discovery.* Cambridge: Cambridge University Press.

Lesh, R., & Fennewald, T. (2010). Modeling: What is it? Why do it? In R. Lesh, C. Haines, P. Galbraith, & A. Hurford (Eds.), *Modeling students' mathematical modeling competencies* (pp. 5–15). New York: Springer.

Mason, J. (1998). Researching from the inside in mathematics education. In A. Sierpinska & J. Kilpatrick (Eds.), *Mathematics education as a research domain: A search for identity,* 2 vols (Vol. 2, pp. 357–378). Dordrecht: Kluwer.

Mason, J. (2002). *Researching your own practice: The discipline of noticing.* London: RoutledgeFalmer.

Mason, J. (2011). Mathematics education: Theory, practice & memories over 50 years. *For the Learning of Mathematics, 30*(3), 3–9.

Mason, J., Johnston-Wilder, S., & Graham, A. (2005). *Developing thinking in algebra.* London: Sage (Paul Chapman).

Mason, J. Burton L. & Stacey K. (1982–2010). *Thinking mathematically*, London: Pearson.

Marton, F. & Booth, S. (1997). *Learning and awareness*. Hillsdale, USA: Lawrence Erlbaum.

Mellin-Olsen, S. (1987). *The politics of mathematics education*. Dordrecht: Reidel.

Niss, M. (2003). Applications of mathematics '2000'. In D. Coray, F. Furinghetti, H. Gispert, B. Hodgson, & G. Schubring (Eds.), *One hundred years of l'enseignement mathématique: Moments of mathematics education in the twentieth century*. Proceedings of the EM–ICMI symposium (pp. 271–284). Geneva: L'Enseignement Mathématique.

Open University. (1980). *PME233 mathematics across the curriculum* (A distance learning course). Milton Keynes: Open University.

Open University. (1982). *EM235: Developing mathematical thinking* (A distance learning course). Milton Keynes: Open University.

Piaget, J. (1971). *Biology and knowledge*. Chicago: University of Chicago Press.

Pólya, G. (1945). *How to solve it: A new aspect of mathematical method*. Princeton: Princeton University Press.

Pólya, G. (1962). *Mathematical discovery: On understanding, learning, and teaching problem solving* (Combined ed.). New York: Wiley.

Sawyer, W. (1959). *A concrete approach to abstract algebra*. London: Freeman.

Schoenfeld, A. H. (1992). Learning to think mathematically: Problem solving, metacognition, and sense-making in mathematics. In D. Grouws (Ed.), *Handbook for research on mathematics teaching and learning* (pp. 334–370). New York: Macmillan.

Silver, E. (Ed.). (1985). *Teaching and learning mathematical problem solving: Multiple research perspectives*. Hillsdale: Lawrence Erlbaum Associates.

Silver, E. (1994). On mathematical problem posing. *For the Learning of Mathematics, 14*(1), 19–28.

Silver, E., & Stein, M. (1996). The QUASAR project: The "revolution of the possible" in mathematics instructional reform in urban middle schools. *Urban Education, 30*(4), 476–521.

Tahta, D. (1972). *A Boolean anthology: Selected writings of Mary Boole on mathematics education*. Derby: Association of Teachers of Mathematics.

van den Heuvel-Panhuizen, M. (1994). New chances for paper-and-pencil tests. In L. Grugnetti (Ed.), *Assessment focussed on the student – L'évaluation centre sur l'élève*. Proceedings of CIEAEM 45 (pp. 213–222). Calgari: University of Calgari.

Watson, A. (2007). *Raising achievement in school mathematics*. Maidenhead: Open University Press.

Watson, A., & Mason, J. (2005). *Mathematics as a constructive activity: Learners generating examples*. Mahwah: Erlbaum.

Chapter 5
Problem Solving, Exercises, and Explorations in Mathematics Textbooks: A Historical Perspective

João Pedro da Ponte

Abstract This paper analyzes the tasks proposed in several Portuguese mathematics textbooks from the nineteenth to the twenty-first century. A look at the nature and intended purpose of these tasks raises interesting issues about school mathematics teaching and learning. Has the meaning of terms such as "problem" and "exercise" been always the same? What other terms have been used in textbooks to designate mathematics tasks? What were the reasons for the changes? The analysis of the evolution that occurred in the terminology as well as in the nature of the tasks proposed to the students provides elements to reflect about what are the changes that have occurred in mathematics teaching and learning and how some changes are more apparent than real.

Keywords Tasks • Problem solving • Explorations • Algebra • Textbooks

Introduction

History stimulates a broad view of social processes and a deep understanding of educational questions. As a secondary school student I always maintained a high interest in historical issues which I kept and developed as a mathematics educator. When I took a position at the University of Lisbon, in 1980, after 6 years as a secondary school mathematics teacher, I had some difficulty in figuring out my own academic identity. As the single mathematics educator at the university, I had no colleagues to discuss the content and meaning of my new job; I looked at recent history striving to understand what had happened in the former period in the field of mathematics education, both in the international setting and in my own country. At the time, the major influence in the curriculum was modern mathematics. Therefore, some questions that I struggled with were: What provided the impetus for this

J.P. da Ponte (✉)
Instituto de Educação, Universidade de Lisboa, Lisbon, Portugal
e-mail: jpponte@ie.ulisboa.pt

© Springer International Publishing Switzerland 2015
E. Silver, C. Keitel-Kreidt (eds.), *Pursuing Excellence in Mathematics Education*, Mathematics Education Library, DOI 10.1007/978-3-319-11952-6_5

movement? What were its main ideas? How those ideas did get translated in the curriculum and in school mathematics teaching and learning? What led this movement to decline? As a graduate student at the University of Georgia, shortly after, I become a student of Jeremy Kilpatrick who participated in this period and could present a firsthand account of the events and the forces behind them.

Besides modern mathematics, I was always intrigued by what happened before. The paper by Jeremy Kilpatrick (1992), about the history of research in mathematics education in the USA, not only provided a look going back much earlier, but it also provided a very comprehensive view the elements that contributed in the development of our field. This helped me to overcome my fragmented view, understanding the relations of mathematics education with mathematics and psychology, its place at the university, the emergence of the professional domain of mathematics education, and its consolidation as an academic community.

As a secondary school mathematics teacher and later as a mathematics teacher educator, I also developed a high interest in problem solving. I was always fond of investigating mathematics issues, striving to understand relationships at a deeper level. As a teacher educator, I addressed questions such as what were the fruitful problems to propose to students at different school levels and how to work them in the classroom (e.g., Ponte and Abrantes 1982). The reading of the paper by Stanic and Kilpatrick (1989), with a historical perspective on problem solving in the mathematics curriculum, was an enlightening experience. I could see the important role of problems in school mathematics, from ancient to contemporary times. And I also could see how an insightful analysis of the thinking of two outstanding scholars – John Dewey and George Pólya – may provide the ground for a deeper understanding of different ways of using problems in teaching and learning.

As a graduate student working with the supervision of Jeremy Kilpatrick, I learned a lot from him in many fields, such as research methods, student learning, assessment, curriculum theory, and teacher education. Perhaps the most important learning concerned the ways of framing questions in mathematics education and pursuing disciplined inquiry into them, considering different viewpoints, weighting evidence, and framing conclusions in a clear and convincing way.

The role of algebra in the school curriculum is another object of attention of Jeremy (Kilpatrick and Izsák 2008). In the 2000s, I become increasingly interested in the teaching of algebra. In Portugal, as in other countries, after the strong algebraization of the curriculum in the modern mathematics period, this topic just disappeared as a central curriculum strand. Of course, there were still equations and functions in the curriculum, but algebra was no longer a main organizing idea of the curriculum – and the students did not even come across with this term. Besides studying the emergent literature on algebraic thinking and early algebra, I took a look on the teaching of algebra in former periods. In a first paper (Ponte 2004) I sought to identify aspects that changed in the approach of first-degree equations in Portuguese textbooks in the period of a century. For that purpose I took a textbook from the end of the nineteenth century, other from the middle of twentieth century, other from modern mathematics (the 1970s), and one contemporary (the 1990s). The analysis shows an interesting evolution driven by the ideas of "starting earlier" and "simplifying," providing room for teaching other topics. With modern mathematics the notation and approach had a significant change, but afterwards, with some additional simplification, they went back to something quite similar to what was before. In another paper (Ponte et al. 2007), I made a similar analysis regarding how second-degree equations are presented in seven school mathematics Portuguese textbooks. The analysis also shows that the approach is increasingly simplified in its content, in the tasks proposed to the students, and in the language. Doing this analysis I noticed interesting contrasts in the tasks proposed to the students at the end of the chapters. Therefore, in this paper, I choose to combine several of these strands – history of mathematics education, problem solving, algebra, and student learning – and discuss the evolution of mathematics tasks proposed in mathematics textbooks.

A Framework to Analyze Mathematics Tasks

Pólya (1957) makes a distinction between routine and nonroutine mathematical problems. In his view, a routine problem (or "exercise") can be solved by using a general method, given in abstract terms or by practicing a variety of examples. In contrast, a nonroutine mathematical problem (or just "problem") requires the use of some judgment or inventive faculty. More recently, Stein and Smith (1998) present another framework for mathematical tasks based on the presupposed cognitive complexity and also in the possible establishment of connections. They classify tasks as having low or high cognitive demand, the first including "memorization tasks" and "procedures with no connections" and the second including "procedures with connections" and "doing mathematics." The framework of the PISA study (OCDE 2004) refers to three different kinds of tasks, also classified according to their level of cognitive demand – reproduction, connection, and reflection. However, we must note that it is impossible to classify a task just by itself, since its meaning is always relative to the person that is to solve it. In fact, a task may be a difficult problem for a given person and just a simple exercise for another person. When we speak of the cognitive demand of a task, we always have in mind the "typical" student to whom it will be proposed.

Besides familiarity with a solution method and cognitive demand, there are other dimensions in which tasks may be classified. For example, the context may be purely mathematical or framed in terms of other school subjects or experiential fields. In this regard, Skovsmose (2001) makes an interesting distinction between tasks that have a "real-life" reference, which involve the real data collected from experience, and "semi-reality," in which the statement corresponds to an abstracted, simplified, and often stereotyped situation that is framed just for school purpose. Another distinction that may be made regards the nature of the representations involved. For examples, the task may be framed in a very symbolic and abstract way with almost no natural language, may require extensive interpretation of text, may involve other representations as images or diagrams, and may require the use of several representations and possibly of conversion of representations.

Fig. 5.1 Different kinds of
tasks, in complexity (*vertical
dimension*) and structure
(*horizontal dimension*)

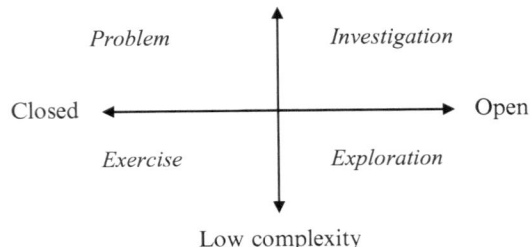

Exercises, explorations, problems, and investigations may be regarded as different kinds of mathematical tasks according to the degree of structure and to the level of cognitive demand – which I prefer to refer to as degree of challenge (Ponte 2005). The exercise is, by far, the most characteristic task in mathematics. Pólya (1957) made a strong case for the idea that, besides exercises, students should also have the opportunity to get involved in solving problems. His distinction between exercise and problem become a key idea in mathematics education. As a consequence, in many countries, everything that is not seen as an exercise is often called a "problem." However, the interest in considering different tasks with specific purposes suggests the need for a refinement of the language. The terms "investigation" and "exploration" are increasingly being used to name tasks that have some element of openness in the aim, the givens, and/or the possible interpretations. It is interesting to note how often these two terms appear in professional journals such as the *Mathematics Teacher* and *Teaching Mathematics in the Middle School* as well as the decision of the NCTM to rename the *Students' Math Notes* as *Students' Explorations*. Similarly to problem solving, when students are involved in exploring and investigating mathematics, they do not have an immediate strategy to use and the first step is to make sense of the task in order to design a strategy to deal with it. Sometimes the two terms are used interchangeably, but often "investigations" carry the idea of being more complex and "explorations" more accessible (Fig. 5.1).

In this paper I focus the attention in the degree of challenge of tasks as well as to the underlying context, the language and representations that are used to frame them, and the thinking the processes involved. And I also look to the way tasks themselves are presented in textbooks.

Mathematical Tasks in Algebra Textbooks of Different Periods

In this section, I look at six different algebra textbooks, from different periods. In every textbook I pay special attention to the tasks proposed to the students related to the topic of second-degree equations. The focus on a single topic is useful to provide a contrast on the dimensions that we look at.

Augusto José da Cunha Cunha (1887): **Elementos de Álgebra (Elements of Algebra***)*

This textbook, with 338 pages, written by a higher education professor (Polytechnical School), aimed at 13–14 years old lyceum students. It is divided in five "books," with book III devoted to second-degree equations (52 pp.). This book includes four chapters, dealing with second-degree roots (16 pp.), second-degree equations with one unknown (26 pp.), equations that may be reduced to the second or the first degree (14 pp.), and second-degree problems (10 pp.). At the end of each chapter, there are exercises and solutions.

Chapter IV is devoted to solving three problems, one from geometry (dividing a line in two parts in some ratio) and two from physics (speed of sound in the context of a well and finding a point illuminated with equal intensity by two bulbs). The problems are first translated from natural to algebraic language, yielding second-degree equations, then solved with the procedures described in the previous chapters, and finally discussed.

At the end of the first three chapters, there are exercises such as: "Solve the following equations"; "Decompose the trinomial in two first-degree factors"; "Find the square root of the polynomial"; "Simplify the expression"; and "Prove the following equalities" (Fig. 5.2). All exercises proposed, starting with exercise 1, are highly structured and complex. At the end of chapter IV, the textbook presents eight problems to solve (Fig. 5.3): four geometrical, two numeric (involving relationships among different numbers), one arithmetic sequence, and another a daily life situation. Solutions are given for almost all exercises.

EXERCICES

Solve the following equations :

I. $\frac{2x-3}{x-3} = \frac{5x-4}{x} - \frac{3}{2}$

II. $\frac{1}{x-1} + \frac{1}{x-2} = \frac{1}{x-3}$

(…)

XIV. $\frac{x^2}{\sqrt{a}+\sqrt{b}} - (\sqrt{a} - \sqrt{b})x = \frac{\sqrt{a^3 b^3}}{\sqrt{a^2 b}+\sqrt{ab^2}}$

XVI. Decompose the trinomial in two fist degree factors

$3x^2 + \frac{3a^2 - 4b^2}{2ab}x - 1$

Fig. 5.2 The first two and two of the last exercises of chapter II

EXERCICES

I. Find five numbers in arithmetic progression so that their sum is 35 and their product is 10395.

VIII. Two asylums distributed, each one, 120$000 reals of donations. The first assisted 40 poor man more than the second; however, this one gave to each poor man 500 reals more than the first. Haw many poor man were assisted by each asylum?

Fig. 5.3 The first and the last problems of chapter IV (indicated as "Exercises")

In this textbook "exercises" are all mathematical questions proposed for the student to solve at the end of each chapter – all with a high degree of difficulty. In some chapters (II and III) the representation is heavily symbolic (algebraic) with no figures and with a minimum of natural language. The thinking processes required are calculations and translations from natural language to symbolic form (in chapter IV). "Problems" are questions that involve a statement in natural language and may be regarded as just a special kind of exercise.

Eduardo Ismael dos Santos Andrea (1924): Compêndio de Álgebra (Algebra Compendium)

This textbook, with 173 pages, was used by 15–16-year-old students. The author was both a university professor (Faculty of Sciences of University of Lisbon) and a high school teacher (Lyceum Pedro Nunes). Chapter VIII has two sections, one on the function $y = ax^2 - bx + c = 0$ and second-degree equation in one unknown (6 pp.) and another on properties of the second-degree trinomial (13 pp.). Chapter IX deals with second-degree problems and their discussion (3 pp.). This small chapter begins with a paragraph noting that "it is necessary to recognize if the roots of the equation can be solutions of the problem" (p. 110). Then, it solves three problems, one involving daily life (distribution of money by different people), another on geometry (revolution cone), and the third about physics (launching a body upwards). Chapter X studies equations "which solution reduces to a 2nd degree equation" (p. 117), bi-squared and irrational equations.

Chapter IX ends with exercises related to it and to the previous chapter with solutions (4 pp.). The exercises proposed are varied, including questions such as: "Solve the equation" (1–21); "Write the equations with the following roots" (22–25); "Discuss, a priori, the equations" (26–29); "Solve the inequality" (33); and others (Fig. 5.4). The set of exercises ends with two problems, both geometrical, one involving a circle and a square and another a sphere and a cylinder (Fig. 5.4). It is interesting to note the last somehow unstructured part of the problems' statement ("discussion").

As in the previous textbook, all tasks proposed at the end of the chapter are called exercises, no matter their difficulty level; the representations are algebraic and in

---

EXERCICES

Solve the equations:

1. $2x^2 - 3x - 2 = 0$

2. $x^2 + \frac{7}{20}x - \frac{3}{10} = 0$

(...)

19. $\frac{1}{a} + \frac{1}{b} + \frac{1}{x} = \frac{1}{a+b+x}$

20. $\frac{(a-x)^2 - (x-b)^2}{(a-x)(x-b)} = \frac{4ab}{a^2 - b^2}$

(...)

35. Given a circle of radius R and a line in its plane at a distance d from its center, it is sought to construct a square with side $2x$, in which one of the sides is a chord of the circle, and which opposite side is in the given line. Discussion.

36. Cut and hemisphere of radius R by a plane parallel to its base, so that the spherical segment of one base obtained has a volume equal to the cylinder which base is the section and which high is the distance of the parallel plans. Discussion.

Fig. 5.4 First two tasks, two additional tasks, and last two tasks from chapter IX

natural language with no figures. Those which statement involves conditions in natural language that must be translated in mathematical language are called "problems." Most of the exercises are quite complex and highly structured, requiring calculations. However, in contrast with the 1887 textbook, this one includes two relatively simple exercises (the first two) and one that may be regarded as less structured (requiring the "discussion" of an equation).

J. Jorge G. Calado (1960, First Edition 1952): Compêndio de Álgebra (Compendium of Algebra)

The textbook is aimed at lyceum students 12–14 years old and has 419 pages. The author is a teacher in the Normal Lyceum Pedro Nunes. There are two chapters regarding second-degree equations and problems. Chapter XX, with 25 pages, has three sections dealing with numerical equations (3 pp.), algebraic solutions (16 pp.), and literal equations (2 pp.), and ends with exercises and solutions (5 pp.). Chapter XXI, with 8 pages, is devoted to second-degree problems. It includes a section on literal problems – conditions of possibility (4 pp.) and a set of exercises and solutions (3 pp.).

Chapter XX closes with 8 exercises (with many subquestions) and solutions, including solving second-degree numerical and literal equations, first without using the quadratic formula and after using it (Fig. 5.5). All exercises are highly structured,

---

EXERCICES

1. Solve the following equations without using the quadratic formula:

a) $x^2 - 9 = 0$ *b)* $4y = 25$ $x^2 - 9 = 0$ *c)* $3x^2 - 12 = 0$ *d)* $7x^2 = 0,9583$ (...)

8. Solve in order to x the following equations:

$(...)j) \frac{a}{x+a} + \frac{a}{x} + \frac{1}{6} = 0$ *l)* $\frac{ax}{x^2-a^2} = 1 + \frac{x}{2x-2a}$

Fig. 5.5 First and last exercises of chapter XX

EXERCICES

Solve the following problems:

I) The sum of the inverses of two consecutive whole numbers is equal to $\frac{15}{56}$.

Calculate the numbers *(National Exam 1939)*

(...)

XXXII) The difference between the squares of the diagonals of two rectangles with the same high is equal to s^2. Calculate the basis of those rectangles knowing that they are for each other as 5 is for 3.

Fig. 5.6 First and last problems of chapter XXI

from simple to a remarkable complexity. Chapter XXI ends with 33 problems, with varying difficulty, and solutions (Fig. 5.6).

The representations are algebraic and natural language with no figures. All chapter XX exercises are mathematical, involving solving equations. The exercises at the end of chapter XXI are problems with numerical, geometrical, or daily life situations. The processes required are calculations and translations. This textbook has much more exercises than the previous ones and, most specially, also a great increase in the proportion of the simpler tasks.

António de Almeida Costa, Alfredo Osório dos Anjos, and António Augusto Lopes (1974): Compêndio de Matemática (Mathematics Compendium)

This textbook, with 280 pages, was written by three lyceum teachers to be used by 14-year-old lyceum students. The topic of second-degree equations is presented after the study of logical notions (such as terms and propositions, expressions, conjunction of conditions), the order relation in IR and inequalities. Section 11 deals

Exercises

(set 1) Solve the following equations:

a) $x^2 - \sqrt{3}x = 0$ b) $x^2 = 4x$ c) $3^2 + \sqrt{3}x = 0$ d) $5x^2 - 1.4x = 0$

(...)

(set 5) Solve the following equations:

(...) c) $4x^2 + 9 = 0$ d) $x^2 - \frac{5}{6}x + \frac{1}{6} = 0$

Fig. 5.7 First and last sets with exercises in section 11

12.5 Solve the following problems:

1) Find to consecutive whole numbers, knowing that the sum of the squares is 61.

...

5) The area of a room may be covered by 3000 squared tiles or with 2500 rectangular tiles. Find the dimension of the rectangular tiles knowing that length surpasses width in 2 cm and this is equal to the squared tiles.

Fig. 5.8 Proposed problems in section 11

with numerical second-degree equations (12 pp.) and section 12 with second-degree problems (2 pp.).

Section 11 begins with solving a numerical problem ("I think of a number..."). The problem leads to writing an equation that is solved using the "zero-product property." Next, it presents a new problem and a new equation to solve and then the equation $\left(2x^2 + \sqrt{2}x = 0\right)$ is solved using such property. The first three points close with four exercises to solve (Fig. 5.7). Point 4 includes two new equations ($x^2 - 9 = 0$ and $4x^2 - 3 = 0$), solved by using special cases of factoring polynomials, and four more exercises. Point 5 solves complete second-degree equations using special cases of factoring polynomials, and end with three exercises. The next two points present more examples of solved equations, of increasing complexity, and end with four exercises, all quite simple (Fig. 5.7). Point 13 presents the canonical form and the notion of incomplete equation showing how to solve them. The quadratic formula is proved by completing the square and with no use of natural language. Finally, this point presents two solved examples and four exercises to solve. In section 12 the textbook solves four problems by several processes (not always by applying the quadratic formula). The first is a numerical problem, the second involves ages, the third is about buying pencils and notebooks, and the fourth is a geometry problem involving a triangle rectangle, highs, legs, and hypotenuse. The section closes with five problems to solve and their solutions (Fig. 5.8).

In this textbook there are few exercises. They are highly structured, requiring calculations or translations, and their difficulty level is low (figures 24–28). The proposed problems are regarded as tasks framed natural language and involve situations from geometry and daily life. An important innovation is that tasks are proposed as the topics are addressed and not at the end of the chapter, as in the former textbooks. Another difference is that problems are no longer presented as a special kind of exercise but as a different task.

António de Almeida Costa, Alfredo Osório dos Anjos and António Augusto Lopes (1983): Matemática Jovem (Young Mathematics)

This textbook, written by the same authors, has 327 pages and aimed at 14-year-old students. Chapter 5, with 15 pages, addresses second-degree equations and problems. The chapter is divided in four parts, beginning with second-degree equations in IR (8 pp.) and second-degree problems (2 pp.) followed by several sets of tasks as well as their solutions.

The first part of the chapter begins by solving a problem from which the second-degree equation is defined. Next, it solves different kinds of incomplete second-degree equations and then the complete equation factoring special cases and finally presents and proves the quadratic formula and applies it in two exercises. In the last point it synthetizes the several kinds of second-degree equations and its solution and possibility in IR, indicating the simplified quadratic formula. The second part concerns second-degree problems and presents four solved problems, two numerical, one involving relations between ages, and the last one on high, legs, and hypotenuse of a right triangle. In the third part, under "complementary activities," the textbook proposes solving nine sets of tasks involving equations and five problems and indi-

---

1. Solve the equations in R:

a) $x^2 - 49 = 0$ (...)   *c)* $x^2 = 63$ *d)* $x^2 + 9 = 0$

(...)

8. Solve, in R, the following equations: (...)

i) $\frac{6}{x} - 5 = x$ *j)* $x - \frac{3}{x-1} = 3$

9. Verify that:

a) $x^2 + (x + 2)^2 = 10^2 <=> x = 6 \lor x = -8$

b) $x(2x - 1) = (x + 1)^2 + 3 <=> x = 4 \lor x = -1$

Fig. 5.9 Fist and last exercises about the 2nd-degree equation

cates their solutions. One must note the complexity of the equations presented in the two last points of task 8 and the use of logical symbols in task 9 (Fig. 5.9).

The chapter ends with a section named "review activities" constituted by questions that involve topics formerly dealt with such as inequalities, systems of equations, factoring polynomials, solving equations, solving problems, and simplifying radicals. The number of tasks proposed to the student to solve increased a lot in relation to the previous textbook and is again put at the end of the chapter – not as simply a list, but as a section on its own right. The tasks are highly structured, framed in algebraic and natural language with no figures, and requiring calculations or translations. One must note that two of the equations proposed have some difficulty, with the unknown in the denominator. The problems are the same as in the previous textbook. An interesting feature of this textbook is that the term "exercise" does not appear, being replaced by the term "activity."

Maria Augusta F. Neves, Luís Guerreiro, and Armando Neves (2004): **Matemática 9 (Mathematics 9)**

This textbook was published in two volumes and aims 14-year-old basic education students. One author teaches in a higher education polytechnic and two are secondary school mathematics teachers. The first volume (128 pages) is divided in four chapters, one dealing with second-degree equations (22 pp.). This chapter has four sections, all with 4 pages: (i) Operations with polynomials. Special cases of factoring polynomials. Decomposition in factors; (ii) Solving incomplete second-degree equations. Zero-product property; (iii) Solving complete second-degree equations. Quadratic Formula; and (iv) Solving second-degree problems. Half of each section presents an explanation of the topic, several solved examples of related questions, and a small synthesis and the other half includes a set of "problems," the last one presented as "reflection/discussion." The chapter ends with a review section named "keywords/knowledge and specific capacities" and another on "evaluation."

The fourth section concerns solving second-degree problems. It suggests drawing a diagram to help writing an equation that relates data and unknown. Three problems are solved, two geometrical (one involving an enlargement of area and the other the fencing of a preschool garden) and one numerical (The product of a number by its third part is 48. What is the number?). The sections end proposing problems to solve, one numerical, four geometrical, two concerning functions, and one riddle.

The evaluation section has two kinds of tasks – five "multiple choice questions" (2 pp.) and nine "development questions", including both equations and problems to solve (Figs. 5.10 and 5.11). Most of the tasks are of low complexity. There is no mention to "exercises," but, in contrast, the term "problem" is very frequent. At the end of the textbook, solutions for all tasks are provided.

In this textbook, there are tasks to solve in each section and at the end of the chapter. Most tasks are highly structured and require calculations and translations,

2. Solve each of the following equations without making use of the quadratic formula.

2.1 $x^2 - 9 = 0$ 2.2 $(x - 5)^2 = 0$

2.3 $-x^2 + 7 = 0$ 2.4 $2x^2 = -\frac{1}{2}x$

2.5 $(x + 3)^2 = 0$ 2.6 $(x - 2)^2 = 9$

2.7 $(x - 4)^2 = 5$ 2.8 $x^2 - 2x + 1 = 16$

3. Observe the figure.
The volume of the parallelepided that the derrick
is pulling is 24 m².
3.1 Find the dimentions of the parallelepided.
3.2 What is the mass of the parallelepided if 5 m²
of the same material have a mass of 7500 kg?

4.For each of the following equations, write it in canonical form and use the quadratic formula
to find the solutions.

4.1 $2x^2 + 3x = -2$ 4.2 $x(x + 2) = 8$

4.3 $x^2 - \frac{1}{3} = -\frac{2}{3}x$ 4.4 $\frac{x-2}{2} - x(x + 1) = -19$

Fig. 5.10 Three of the tasks of higher complexity at the end of the chapter

9. The space, in meters, travelled by a body in free fall and with no friction is, approximately, $e = 5t^2$ where t represents the time, in seconds, since the beginning of the fall.

9.1 A coin was dropped from the top a tower and took 3 seconds to reach the ground. What is the high of the tower?

9.2. A parachutist fell from an airplane in free fall for 720 m. How much time he took to open the parachute?

Fig. 5.11 Last task at the end of the chapter

and some also require understanding concepts and interpretation. They are framed in algebraic and natural language with frequent figures and some are purely mathematical (numerical and geometrical) and others draw on daily life situations. One must note that most of the tasks are of low complexity, but they are most often termed as "problems."

Conclusion

In all manuals, the study of second-degree and second-degree problems are associated, either in the same chapter or in contiguous chapters. The main representations are algebraic and natural language; only very recently figures emerged with an important role. Throughout this period, the most common problems are numeric, geometric, and daily life, with one or another physics problem. Besides the discourse, we do not see much evolution on these problems, except in the daily life problems, in which we note a remarkable change of themes.

In this set of textbooks, there is a very interesting development in the tasks proposed for the students to solve. The questions presented decrease progressively in their complexity and, in addition to calculation and translation of natural to symbolic language, they begin also requiring some understanding of concepts and interpretation of situations. Another important evolution concerns the place of tasks in textbooks – from the short list at the end of the chapter to the long list and finally to its scattering through the chapter.

It should be noted, finally, the resignification process that the terms "exercise" and "problem" assumed over time. At first, the exercises are very complex tasks and problems are a particular type of exercises, with a statement in natural language. Over time, the exercises begin including a wide variety of tasks that, in general, assume less and less complexity. Later, the term "exercise" disappears and is replaced by "activity" and, in the last manual, the more inclusive term that means any type of task is "problem." In this evolution, we note an effect of mathematics education, which devalues the concept of exercise (contributing to its disappearance in textbooks) and values the concept of the problem (contributing towards its visibility).

The tendency to simplify the tasks proposed to students is a consequence of changes in the role of the school, in particular its universalization at increasingly higher levels. The broadening of representations used and the processes required in students is certainly due to a better understanding of their learning processes. But, above all, it is worth mentioning the great capacity of adaptation of textbooks, which appropriate terms are used in mathematics education, such as "activity" and "problem," resignifying them according to their own logic.

References

Andrea, E. I. S. (1924). *Compêndio de Álgebra*. Lisboa: Imprensa Nacional de Lisboa.
Calado, J. J. G. (1960). *Compêndio de Álgebra*. Lisboa: Livraria Popular de Francisco Franco.
Costa, A. A., Anjos, A. O., & Lopes, A. A. (1974). *Compêndio de Matemática*. Porto: Porto Editora.
Costa, A. A., Anjos, A. O., & Lopes, A. A. (1983). *Matemática Jovem*. Porto: Porto Editora.
Cunha, A. J. (1887). *Elementos de Álgebra* (5th ed.). Lisboa: Livraria de António Maria Pereira.
Kilpatrick, J. (1992). A history of research in mathematics education. In D. A. Grouws (Ed.), *Handbook of research on mathematics teaching and learning* (pp. 3–38). New York: Macmillan.

Kilpatrick, J., & Izsák, A. (2008). A history of algebra in the school curriculum. In C. E. Greenes & R. Rubenstein (Eds.), *Algebra and algebraic thinking in school mathematics* (pp. 3–18). Reston: NCTM.

Neves, M. A. F., Guerreiro, L., & Neves, A. (2004). *Matemática 9* (1st ed.). Porto: Porto Editora.

OCDE. (2004). *Learning for tomorrow's world: First results from PISA 2003*. Paris: OCDE.

Pólya, G. (1957). *How to solve it: A new aspect of mathematical method* (2nd ed.). New York: Doubleday.

Ponte, J. P. (2004). As equações nos manuais escolares. *Revista Brasileira de História da Matemática, 4*(8), 149–170.

Ponte, J. P. (2005). Gestão curricular em Matemática. In GTI (Ed.), *O professor e o desenvolvimento curricular* (pp. 11-34). Lisboa: APM.

Ponte, J. P., & Abrantes, P. (1982). Os problemas e o ensino da Matemática. In *Ensino da Matemática: Anos 80* (pp. 201–214). Lisboa: SPM.

Ponte, J. P., Salvado, C., Fraga, A., Santos, T., & Mosquito, E. (2007). Equações do 2.º grau do fim do século XIX ao início do século XXI: Uma análise de sete manuais escolares. *Quadrante, 16*(1), 111–145.

Skovsmose, O. (2001). Landscapes of investigation. *ZDM, 33*(4), 123–132.

Stanic, G. M. A., & Kilpatrick, J. (1989). Historical perspectives on problem solving in the mathematics curriculum. In R. I. Charles & E. A. Silver (Eds.), *The teaching and assessing of mathematical problem solving* (pp. 1–22). Reston: NCTM e Lawrence Erlbaum.

Stein, M. K., & Smith, M. S. (1998). Mathematical tasks as a framework for reflection: From research to practice. *Mathematics Teaching in the Middle School, 3*(4), 268–275.

Chapter 6
From Mathematical Problem Solving to Geocaching: A Journey Inspired by My Encounter with Jeremy Kilpatrick

Thomas Lingefjärd

Abstract The activity of problem solving is probably as old as mankind. Within mathematics education, there is a vast amount of books, textbooks, general articles, and research articles published concerning the use of and learning of problem solving. There are many different views and opinions regarding what a problem really is, and sometimes problems are divided into closed or open problems. An open-ended problem is a problem that has several or many correct answers and several ways to the correct answer(s). One activity developed by people using new technology together with open-ended problems is the activity geocaching. In this paper, I will try to give a flavor of how intriguing and almost addictive mystery solving in geocaching might be.

Keywords Problem solving • Technology • Informal learning • Geocaching

Meeting Jeremy

The seventh ICME conference took place in Quebec in the summer of 1992. It was the first ICMI conference I attained and one of my colleagues from the University of Gothenburg nodded towards a person at the other side of the room during a coffee break and said: That is Jeremy Kilpatrick. I will try to invite him to come for a visit at our department next spring. I was not aware of it then, but that visit in fact changed my life.

Next spring Jeremy came to our department and since I was the head of the mathematics education group, my responsibility was to equip Jeremy with a computer. We had just received a new Macintosh computer with a large monitor and I offered Jeremy to sit and work by that computer. Unfortunately the computer only understood Swedish (this was many years before you could select the language of the operative

T. Lingefjärd (✉)
Mathematics Education, University of Gothenburg,
Department of Education Box 300 SE, 405 30 Göteborg, Sweden
e-mail: Thomas.Lingefjard@gu.se

© Springer International Publishing Switzerland 2015
E. Silver, C. Keitel-Kreidt (eds.), *Pursuing Excellence in Mathematics Education*, Mathematics Education Library, DOI 10.1007/978-3-319-11952-6_6

system on a computer yourself), so during his 2 months at our department, Jeremy learned Swedish expressions for Save as, Open, and Print, just to mention a few.

Jeremy's office was also close to mine and we started to take breaks and have coffee together; the concept of "fika" is very important in daily life in Sweden (literally meaning coffee with something added to it, a cookie or a bun). During those daily conversations and discussions, Jeremy Kilpatrick tried to convince me to visit Athens, Georgia, and that I should apply for a Ph.D. position at Jeremy's department. The problem was that neither I or nor my colleagues actually knew anyone who had done that journey before. When I described this eventually drastic change of our lives to my wife, she did not really buy Jeremy's arguments as easy as I did.

Becoming a Doctoral Student

In the spring of 1994, I did my first visit to Athens, Georgia. I met with Jeremy's students and colleagues; I was guided around the town by Jeremy and his wife Cardee and I stayed at their house at Woodlawn Avenue. I very much liked downtown Athens and started to think about an academic career grounded on edification from UGA. I knew already then that it would cause some domestic arguing and that I would need some convincing arguments, but Jeremy promised to help me and he really did. I joined the Ph.D. program in the fall of 1994, moved to Georgia with my family during the summer of 1995 and graduated in 2000, and have never regretted that adventure. Neither has my wife or our children; we had a wonderful time in the USA and in Georgia. More details about that time can be read in Lingefjärd (2001).

My doctoral work was first aiming at the history of mathematics education; I had and still have a deep interest in history. But my interest for technology and mathematical modeling took over and I remember that Jeremy was just a little bit questioning, but he never expressed a rejecting sentence about my direction. Although technology not was Jeremy's main interest, I believe that he saw that it would become a growing concern for everyone teaching mathematics in the future.

Where Technology Are Inviting Us

Eventually, I found out that Jeremy had a deep interest in problem solving, another important meeting point for us. A lot of problem solving might easily be connected to the use of technology today. But problem solving also has real historical roots. It is a historical fact that one of the oldest mathematics textbooks we humans have in our possession is a collection of mathematical problems and puzzles. This manuscript is called either the Ahmes papyrus, after the Egyptian writer who wrote it down, or the Rhind papyrus, after the Scotsman Henry Rhind, who bought the papyrus during a vacation trip in Egypt in 1858. The Ahmes manuscript from 1650 B.C. was in turn a copy of the original books from King Amenemhet II (a king of Egypt from 1849 to 1801 B.C.). It is essentially a collection of mathematical problems in the domain of brainteasers (see Danesi 2002, p. 5). Many authors throughout history have devoted their work to books about problems and problem solving. One of the bestsellers during the medieval age was a book called The Book of Games (Mohr 1997, p. 11).

During the last 13 years, a new way that combines problem solving and technology is the constantly growing activity geocaching. Geocaching is an outdoor activity (although some caches are hidden in caves or rock shelters) in which the geocachers (participants) use a Global Positioning System (GPS) receiver or other navigational techniques to hide and seek containers (called geocaches or just caches) anywhere in the world. A typical cache is a small waterproof container containing a logbook and sometimes a pen. Larger containers can also contain items for trading, usually small toys or trinkets of little or no value. Geocaching is often described as a "game of high-tech hide and seek," sharing many aspects with orienteering, treasure hunting, and way marking. I prefer to call Geocaching an activity instead of a game. Geocaches are currently placed in over 100 countries around the world and on all seven continents, including Antarctica. As of September 2013, there are 2,229,926 active geocaches and over six million geocachers worldwide. Since not all geocachers list their caches on the major service provided by geocaching.com, there are probably a large number of unrecorded caches around the world as well (see www.geocaching.com).

Geocaching started in May 2000, when the US government turned off the selective availability function, which was an intentional error in the GPS technology. That changed the GPS accuracy from 300 ft to the range of 6–20 ft. Two months earlier, David Ulmer, a computer consultant, had already raised the idea of testing GPS accuracy by hiding a navigational target in the woods. Ulmer called this activity the Great American GPS Stash Hunt and explained the idea as: Hide a container somewhere, post the coordinates on an Internet GPS users' group, and challenge the others in that group to find the container. The day after the selective availability function was switched off, on May 3, 2000, Ulmer placed the very first cache in the woods near Beaver Creek, Oregon, and the activity of Geocaching was born.

Since that day, the activity has grown enormously and consequently it has divided itself into the hiding of, and search for, several different types of caches. The most common of these cache types are as follows (excerpted from more complete descriptions at www.geocaching.com):

- *Traditional cache* is the original cache type. The coordinates listed on the traditional cache page are the exact location for the cache.
- *Multi-cache* ("multiple") involves two or more locations, the final location being a physical container. Most multi-caches have a hint to find the second cache; the second cache has hints to the third, and so on.
- *Mystery cache* or *puzzle cache* is the catchall cache type; this form of cache can involve complicated puzzles that you first need to solve to determine the coordinates. Because of the increasing creativity of geocaching, this cache type has become the staging ground for new and unique challenges.

- *Wherigo cache* is a toolset for creating and playing GPS-enabled adventures in the real world. By integrating a Wherigo experience, called a *cartridge*, with finding a cache, the geocaching hunt can be an even richer experience. Among other uses, Wherigo allows geocachers to interact with physical and virtual elements such as objects or characters while still finding a physical geocache container.
- *Earth cache* is a special place that people can visit to learn about a unique geoscience feature or aspect of our Earth. Earth Caches include educational notes and details about where to find the location (latitude and longitude).

Mystery Caches and Mathematics

There is no way of detecting exactly when or where the first mystery cache was published, but soon after the first regular cache in which you were given the coordinates to find the cache, the first mystery cache, where you have to solve a puzzle to derive the coordinates, was constructed. Such caches are the most rapidly growing part of geocaching, and the variety and creativity in creating mystery puzzles is enormous.

> As these examples show, problems have a long history in the mathematics curriculum. However, primarily within the last century, discussions of the teaching of problem solving have moved from advocating that students simply be presented with problems or with rules for solving particular problems to developing more general approaches to problem solving. (Stanic and Kilpatrick 1988, p. 4)

If Jeremy had known then that geocaching would become so enormously important, he couldn't have phrased it better. A mystery or puzzle cache is constructed this way: You measure the GSP coordinates where you want to hide your cache and use some way to disguise these coordinates. The difficulty is labeled from one to five, and caches with a classification 5/5 indicate both a severe difficult puzzle and severe difficulties to physically get to the cache. An easy and rather simple way to hide the cache coordinates would be as follows: Imagine that you hide your cache at the coordinates of N 59.19.650 E 018.550. With simple number theory, these coordinates could be disguised as 5^5, 9^9, 1^1, 9^9, and since $5^5 = 3,125$, this sequence could be written 3125, 387420489, 1, 387420489, 46656, 3125, 0, and so forth. The directions N (North) and E (East) are often omitted in the puzzle. Now it is up to the geocacher to understand that $3,125 = 5^5$ and so forth. Normally this way to disguise the coordinates would yield a cache of difficulty 1.5 or maybe 2.

There are so many ways that people have invented to disguise the coordinates that it is impossible even to try to give an overview. Different code systems are very popular, images with hidden information, algebraic expressions, geometrical, number theory, probability puzzles, and so forth. But also historical, geographical, or other facts are often used. Some puzzles are long stories in which the information is hidden in the text in an implicit way. As a person with a mathematical background, I find puzzles with mathematics rather easy while some others might be almost impossible to solve. Sometimes this is referred to as "thinking outside the box", sometimes rather hard for people trained in mathematics.

> No longer was it assumed that the study of mathematics inevitably improves one's thinking. This view set the stage for a greater emphasis by mathematics educators on how, exactly, students might improve their thinking ability, or reasoning ability, or problem-solving ability, through the studies of mathematics. Many of our professional forebears, however, were reluctant to give up the tradition going back to Plato that gave such a prominent place to mathematics in the school curriculum. (Stanic and Kilpatrick 1988, p. 11)

From the beginning, there were caches which often could be found by pure luck or by serendipity, but today most mystery caches from level 2 and upward are difficult to decode and difficult to find. But the harder the problems become, the more interest it draws from the surrounding geocaching community.

It is amazing that there is such a deep desire to solve problems produced by a leisure time activity. I have just of curiosity registered how many hits there are on the Web page for the cache when I publish a new cache. Amazingly, there are several hundred hits the first week of the cache. Compare that to the result in any normal class around the world when a new problem is introduced to the students. Many people also do geocaching as part of their ordinary life when going to or home from work or when travelling. When I was in New York for the New York Marathon in early November 2011, I had solved several mystery caches which were placed in Central Park and I managed to find and log 7 of them while I was there. At some of the caches I was logging, I met others who had come to New York in order to go geocaching in Central Park. What is driving me and others to do this?

Theoretical Standpoint

From a theoretical standpoint, Geocaching is the creation of a didactical situation. The cache owner creates a puzzle or mystery that other geocachers of course are free to solve but also feel encouraged by and sometimes almost forced to solve. In his theory of didactical situations in mathematics, Guy Brousseau (1997) states that:

> The modern conception of teaching therefore requires the teacher to provoke the expected adaptation in her students by a judicious choice of "problems" that she puts before them. These problems ... must make the students act, speak think and evolve by their own motivation... The student knows very well that the problem was chosen to help her to acquire a new piece of knowledge, but she must also know that this knowledge is entirely justified by the internal logic of the situation and she can construct it without appealing to didactical reasoning. Not only can she do it, but she must do it because she will have truly acquire this knowledge only when she is able to put it to use by herself in situations which she will come across outside any teaching context and in the absence of any intentional direction. Such a situation is called an adidactic situation... This situation or problem chosen by the teacher is an essential part of the broader situation in which the teacher seeks to devolve to the student an adidactic situation which provides her with the most independent and most fruitful

interaction possible… She is thus involved in a game with the system of interaction of the student with the problems she gives her. This game, or broader situation, is the didactical situation. (pp. 30, 31)

From this standpoint, it is obvious that the cache owner who creates and publishes a mystery cache and the geocachers who are trying to solve the mystery are actors in a specific didactical situation. At the same time, they are also free agents in a vague scene of informal learning. The cache owner of a mystery cache has no way of forcing other geocachers to solve the problem. Nevertheless, in large cities as many as 500 geocachers might try very hard to solve a specific mystery. There are difficult mysteries that can take months to solve. Some geocachers are working together when solving mysteries, but often this interaction is taking place over the Internet and on random time slots. Some geocachers are hunting the glory of FTF (first to find) or STF (second to find); others are merely happy to solve a mystery now and then and not concerned if several other geocachers already solved it. The community is rather secret and reluctant to "give away" solutions, although some geocachers trade solutions between each other. You are not allowed to be a 'leech'; you have to give what you take so in the long run you are expected to publish caches if you log caches. Every cache that is found has a history of comments on its home page, but when successful geocachers log their visit they carefully ensure that no one can understand where the cache is hidden from what flourish story they write in their log.

If you are interested in learning more, register on geocaching.com. It is free. My alias is ThomasLinge on geocaching.com. See you there.

Research on Geocaching

There have been several studies on geocaching as an outdoor activity, which clearly indicates in the direction that geocaching is a growing phenomenon. O'Hara (2008) states:

In sum, the study has shown the importance of looking beyond the simple in situ consumption of a "treasure hunt". Rather it is important to consider it as an ongoing practice that acquires social significance through its positioning within an on-line community. In addition, it is important to consider creation activities as consumptions. We hope this perspective provides more general insight into the understanding and design of location-based experiences. (p. 12)

Buck (2009) defended her dissertation at the University of Alabama entitled: "The Motivational Effects of a GPS Mapping Project on Student Attitudes Toward Mathematics and Mathematical Achievement." Buck's research questions were as follows:

1. Will Geocaching affect students' attitudes toward mathematics?
2. Will Geocaching affect students' math achievement as evidenced on the New Century Mathematics Diagnostic Test?

3. Will there be a gender difference in attitudes toward mathematics?
4. Will there be a gender difference in attitudes toward Geocaching?

Buck (2009) claims that she found a significant increase in boys' and girls' attitudes towards both mathematics and geocaching during her study, which involved one experimental and one control group:

> The results showed a significant difference between the experimental group and the control group. The experimental group had a higher post-treatment mean and a greater pre- to post-treatment gain than did the control group, indicating that the treatment protocol, mathematics-based GPS mapping activities, positively affected students' attitudes toward mathematics. (p. 79)

Furthermore, in her dissertation "Geolearners: Informal Learning with Mobile and Social Technologies," Gill Clough investigated what geocachers thought they learned when doing geocaching. Her Web survey received 661 responses over a period of 3 weeks, 70 % male and 30 % female. The survey was mainly general but also included a small section about puzzle caches. Clough (2009) concludes that solving the challenges devised by the creators of puzzle caches may also result in learning opportunities:

> We think that puzzle caches which require you to search the internet are often excellent for expanding your knowledge. We have often gone beyond the answer required to find out more just for our own interest. (Survey response 211)

> Some "high-tech" caches have required to learn something new in mathematics and information technology. Thanks to pair of caches, I came to know birds and genetics. (Survey response 490) (p. 269).

> I know my children have learned a great deal. While out in nature, I have a great venue in which to teach my children about animal and plant life, history, geology, history, technology, and even mathematics and cryptology. (Survey response 106) (p. 273)

There is a growing interest from the formal learning society, such as schools, to learn how to create learning scenarios with the help of outdoor activities such as geocaching. The fact that you can do geocaching or decide your position with GPS coordinates not only with handheld GPS technology but also with today's so-called smartphone is an interesting challenge for many teachers.

Smartphone

One reason for the growing interest in geocaching around the world is the possibility to perform your geocaching with the aid of your smartphone. The extremely rapid development of smarter telephones together with the capacity of inbuilt GPS technology and the possibility to download new and individual apps creates a fascinating new personal tool which is definitely something much more than just a telephone.

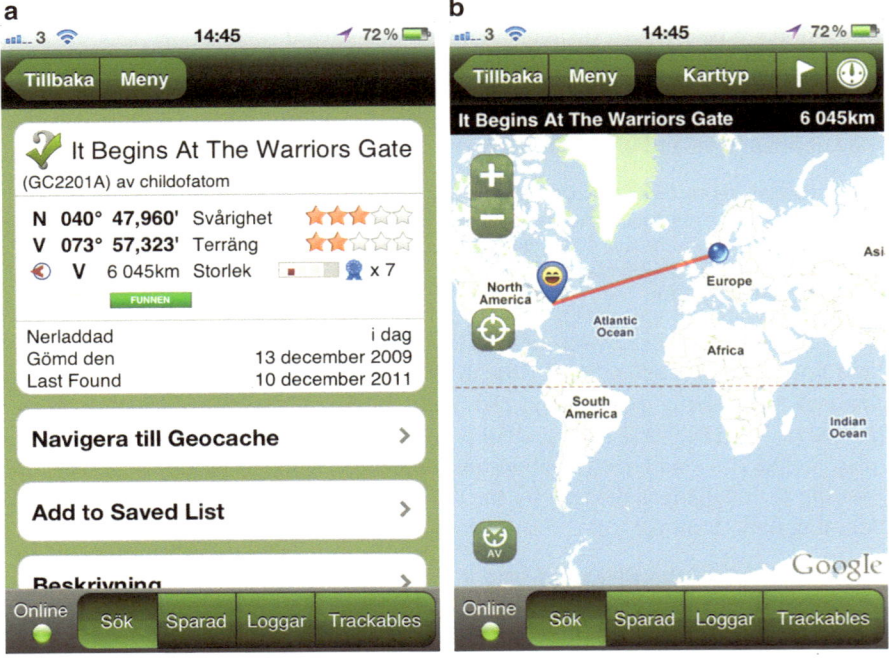

Fig. 6.1 (**a**, **b**) My iPhone as problem solver and as a map

When my telephone guides me to a secret place in the search after a cache I have solved at home and when I am in Central park at Manhattan, then it is not just a telephone. It is a map, a communicative tool, a guide, a sort of safety. It helps me maneuver in a park 6,045 km from home, a park where someone has hidden a secret cache, disguised behind a clever mystery. See Fig. 6.1a, b.

Today I never travel to a new town or place around the globe without checking out the caches that are hidden there. If I have time I start to solve new mysteries months before I go and plan how I should be able to log them when I am there. Jeremy probably would try to frame this activity as a recreational activity, but I see it as something much bigger, something related to us as humans, always intrigued and interested in mysteries.

> *Problem solving as recreation.* The subtheme of recreation is related to that of motivation because student interest is involved, but in the case of recreation, problems are provided not so much to motivate students to learn as to allow them to have some fun with the mathematics they already have learned. Presumably, such problems fulfill a natural interest humans have in exploring situations. The problem shown earlier from the Ahmes Papyrus is a good illustration. (Stanic and Kilpatrick 1988, p. 13)

It seems to me that Jeremy was ahead of all geocacher when he wrote this. If we just change the word students to geocachers, it could have been one of the opening paragraphs of this article. Especially the last two sentences are exactly what I am talking about. It proves that we humans never know what will come. What will be next after smartphones and geocaching?

References

Brousseau, G. (1997). *Theory of didactical situations in mathematics* (N. Balacheff, M. Cooper, R. Sutherland, & V. Warfield, Eds. & Trans.). Dordrecht: Kluwer.

Buck, L. (2009). *The motivational effects of a GPS mapping project on student attitudes towards mathematics and mathematical achievement.* Doctoral dissertation, University of Alabama, Tuscaloosa.

Clough, G. (2009). Geolearners: Informal learning with mobile and social technologies. Doctoral dissertation, Open University, Buckinghamshire.

Danesi M. The puzzle instinct. The meaning of puzzles in human life. Bloomington: Indiana University Press; 2002.

Lingefjärd T. Why I became a doctoral student in mathematics education in the United States. In: Reys R, Kilpatrick J, editors. One field, many paths: U. S. Doctoral Programs in Mathematics Education. Providence: American Mathematical Society; 2001. p. 135–7.

Mohr, M. S. (1997). *The new games treasury: More than 500 indoor and outdoor favorites with strategies, rules, and traditions.* (Enlarged edition of *The games treasury*, 1993). Boston: Houghton Mifflin.

O'Hara K. Understanding GeoCaching practices and motivations. In: Burnett M, Costabile MF, Catarci T, de Ruyter B, Tan D, Czerwinski M, Lund A, editors. Proceeding of the twenty-sixth annual SIGCHI conference on Human Factors in Computing Systems. New York: Association for Computing Machinery; 2008. p. 1177–86.

Stanic G, Kilpatrick J. Historical perspectives on problem solving in the mathematics curriculum. In: Charles R, Silver E, editors. The teaching and assessing of mathematical problem solving. Hillsdale: Lawrence Erlbaum/National Council of Teachers of Mathematics; 1988.

Chapter 7
Ruminations on the Generated Curriculum and Reform in Community College Mathematics: An Essay in Honor of Jeremy Kilpatrick

Vilma Mesa

Abstract In this essay I elaborate on a notion of curriculum, put forward by Jeremy, that highlights the tensions that emerge from the curious interplay between reform, teaching, learning, and culture. I use the setting of American community colleges, to illustrate some of these tensions. I provide vignettes of teaching mathematics at community colleges in different levels of mathematics, to illustrate these tensions, which result in an anticipated stability of the curriculum: the stability is a consequence of the current historical, societal, political, and cultural conditions that surrounds this particular institution.

Keywords Curriculum • Instruction • Community colleges • Curriculum reform

I came to the United States in 1995 to pursue graduate studies in mathematics education at the University of Georgia. Back in Colombia I had been part of "una empresa docente," a research group in mathematics education led by Pedro Gómez at the University of Los Andes in Bogotá. We had been doing some curriculum reform in our department since 1987 (Gómez et al. 1995; Gómez and Fernández 1997; Gómez and Mesa 1995; Mesa and Gomez 1996; Perry et al. 1996a, b). I had met Jeremy two years before when he participated as main speaker at the first International Symposium in Mathematics Education that we organized in Colombia. I was familiar with the NCTM Standards, which had only recently been published in Spanish (National Council of Teachers of Mathematics 1991) when I took a curriculum class with Jeremy. I was ready to learn about using curriculum as a vehicle for reform. It was very exciting. For our first class, Jeremy had assigned us to read the introductory chapter of the *Standards* (National Council of Teachers of Mathematics 1989). To begin the class he asked us to jot down ideas about the

V. Mesa (✉)
School of Education, University of Michigan, 4041 SEB, 610 East University,
Ann Arbor, MI 48109-1259, USA
e-mail: vmesa@umich.edu

© Springer International Publishing Switzerland 2015
E. Silver, C. Keitel-Kreidt (eds.), *Pursuing Excellence in Mathematics Education*, Mathematics Education Library, DOI 10.1007/978-3-319-11952-6_7

question: what is curriculum? I remember being struck by the question. At the time I saw curriculum as the math content that teachers are supposed to teach. I did not volunteer an answer. The discussion that ensued was enlightening. Some students proposed: "the list of the topics you cover, when you cover them, and for how long." "Yes, that could be," I thought, but that was the syllabus for me, and the syllabus could not be a curriculum, an expression of it, maybe. Another student proposed: the courses that students take. Yes, that could be, but I would have called that a program of studies, not necessarily a curriculum. Were we just engaging on a word game? I could have written, "the math students are supposed to learn" but focusing on math content excluded time or sequencing, and, more importantly, whether or not it was meant to be for *all* students—the main theme of the *Standards*. I realized that there were official mandates and also constraints on resources (e.g., time). What about the content that teachers actually teach? I remembered how pressed for time I was teaching the undergraduate math courses and how many times I had to skip sections listed in the syllabus. And then there is the way in which teachers teach. Is that part of the curriculum too? It had been part of the reform, certainly. I panicked. How are we supposed to make reform happen via curriculum if we don't even know what curriculum is? The discussion in that lesson and the rest of the semester also made me realize that decisions about what mathematical content is supposed to be taught or learned are not clear-cut. It highlighted the importance of understanding what happens in the classroom and, moreover, raised questions about what is possible to do in terms of real changes via curriculum.

The following quote, taken from Jeremy's review of *Mathematics Education as a Scientific Discipline* (Biehler et al. 1994), gives us a different way of conceiving curriculum:

> Part of the conceptual difficulty with curriculum content may lie in the view of mathematics as pre-existing somewhere 'out there' in need of being transmuted into an elementary form that teachers and students can encounter it. Maybe the mathematical content of the curriculum might be more productively thought of as *generated* during the process of instruction rather than as prepared and packaged beforehand. At some level, the curriculum, of course, needs to be laid out as a course to be traversed. But at its heart, it is more an intellectual adventure whose details and direction cannot be foreseen. The mathematics prepared for instruction, as enshrined in official documents and text materials, may bear little resemblance to the mathematics of the curriculum encountered by students. (Kilpatrick 1997, pp. 737, emphasis added)

In this chapter I present an essay about what it would mean for curriculum to be studied as "generated during the process of instruction" and the implications of this conception of curriculum for reform. The essay is divided into two parts. In the first part, I present current definitions of curriculum in math education, followed by three vignettes of mathematics instruction at community colleges. I use these vignettes to talk about the generated curriculum in the lessons. In the second part, I expand on the community college context and discuss the possibilities of reform of mathematics education at community colleges via curriculum, given the possible stability of the generated curriculum in this setting. Before, however, I provide a short description of the community college to contextualize this piece.

Community colleges are a very special American higher-education institution offering opportunities for vocational training, degree completion, career advancement, and personal enrichment. Community colleges enroll a significant number of undergraduate students: in 2010, 46 % of all the undergraduate population and nearly 49 % of all enrollments in undergraduate mathematics courses were at community colleges (Blair et al. 2013; Dowd et al. 2006).

A major problem for community colleges is mathematics remediation, that is, preparation of students in the mathematics content that is taught in K-12 education, which would allow students to take college courses. Remediation enrollment at 2-year colleges has increased 20 % in 10 years, from 799,000 in 1995 to 964,000 in 2005, proportionally to the community college enrolment growth and to the decrease of offering of remediation courses in 4-year institutions (Lutzer et al. 2007, p. 137). With failure rates ranging from 40 to 70 % in courses before precalculus (Attewell et al. 2006; Bahr 2008; Waycaster 2001), there are concerns that in spite of their open admissions and affordability, the community colleges are costly institution with strict demands for degree completion, which, paired with the high failure in basic courses, results in deterring students from achieving their original goals (Clark 1960; Melguizo et al. 2008). In addition, in spite of being called "teaching institutions," community colleges in general lack incentives to improve teaching (Grubb 1999). This is the context in which this essay is situated.

Part 1: Various Notions of Curriculum

From the Latin *currere*—to run—the word curriculum referred initially to the course that runners followed in a competition. Today, it has multiple uses: it refers to the sequences of courses that a student can take, the topics that are covered in a given grade, or the content, skills, competencies, and habits of mind that a person needs to acquire through schooling in order participate successfully in the society. The classical distinction between intended, implemented, and attained curricula (Travers and Westbury 1989) helps differentiate states of the mathematics content in the curriculum as it gets proposed and used by different people; it allows for a dynamic view of the content dimension of the curriculum, and in particular the *intended* curriculum corresponds, to some extent, to the idea of *generated curriculum* as articulated in the quote.

These uses of curriculum, however, only seem to be focusing on the *content* of the curriculum and do not recognize other dimensions that play significant roles in defining it, namely, how students learn, how teachers teach, and what the society values as important to learn. Whereas the content (and skills and procedures) of mathematics are important, the different ways in which we believe students learn, or teachers should teach, or what should be assessed, equally define what the curriculum is.

Curriculum, therefore, can be defined as a teaching and learning plan that has four different dimensions—conceptual (content, skills, procedures), cognitive (how

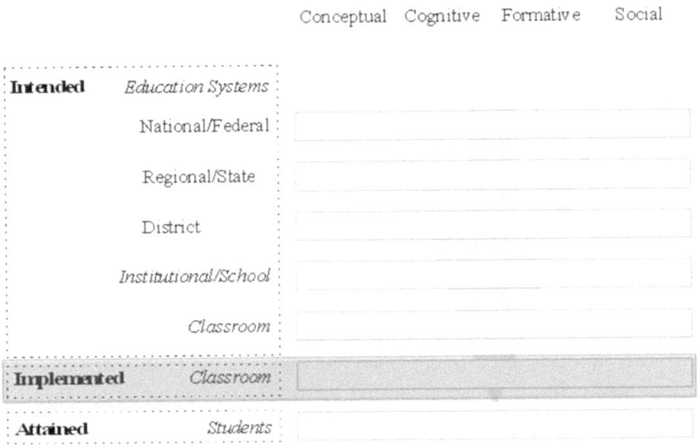

Conceptual Cognitive Formative Social

Fig. 7.1 Dimensions and levels of curriculum. The *arrow* suggests the direction of assumed influences in a curricular system. The *gray area* represents the level at which the generated curriculum can be studied (Adapted from Mesa et al. 2013)

students learn), pedagogical (how teachers should teach), and social (what is valued)—each of which manifests differently at each level of an educational system, at the global, national, institutional, classroom, and student level (Fig. 7.1).

In this definition, the generated curriculum corresponds not only to the mathematical content, as proposed in Jeremy's quote, but also to the ways of learning, pedagogies used and what is valued in any particular class at any particular moment in time as teachers, students, and content come together in the classroom. Documentation of the implemented curriculum over time (e.g., a week, semester, or year) can give an accurate description of the generated curriculum. A description of individual lessons can shed light on the different qualities of this generated curriculum.

The three vignettes that I present next come from different community college instructors, observed in two different years, and representing three different levels and types of courses. As part of a larger project, each instructor was interviewed about their work at their college and their views of mathematics, teaching, learning, and the students they had. Each instructor was observed three times, in a course of their choosing. The teachers represent a range of backgrounds, views of learning, instruction, and values. I discuss the vignettes in light of Jeremy's proposal for seeing curriculum not as "preexisting somewhere" but "generated during the process of instruction." The first vignette comes from a remedial course; the second vignette comes from a probability class, whereas the third one comes from a trigonometry class. I start each vignette with a brief contextualization of the lesson, followed by a short excerpt from the lesson, and concluding with a short analysis of the generated curriculum illustrated by the excerpt.

Basic Mathematics: Converting Percents to Decimals

After affably greeting his 16 students, Erik, opens up his power point presentation for the arithmetic class. It is Thursday 6:00 pm. There are young and middle-aged students. A woman, in her mid forties wears clothing suggesting that she might be an assistant in a doctor's office—colorful front-buttoned, short-sleeved blouse, and hospital-green pants. Another older male is sitting two rows behind wearing a tie and a suit. I recognize a woman in her 20s whom I had met about 6 years ago as a high-school student, who used to dress in Goth style and to hang out in the back of her classes playing with her long hair. Today she looks different, her hair is short, she has a plain t-shirt and wears jeans, and although she is sitting in the back, she is listening intently to the presentation. The lesson will be on conversions between decimals, fractions, and percentages. The following excerpt corresponds to the first five minutes of the class:

Erik: So percent to decimal. [There is a power-point slide with the title, the process, and an example] To convert a percent to a decimal, what we do is we simply move the decimal point to the number, in the number, two places to the left and omit the percent sign. There's really a reason for this and one thing I really want you to keep in mind is that the percent sign is really synonymous, it's not equal but equivalent to saying multiply the number by 1 over 100. Ok. That's really what the percent time means; it's just a notation that we use. Another way of thinking about it is that the percent sign is equivalent if you multiply by .01. Both of those are equivalent statements. So what we do is if we have 58 %, (writes on board 3 seconds) the rule says we move the decimal point in the number two places to the left and omit the percent sign. Well, remember with a whole number, the decimal's at the back of the number. We want to move the decimal two places to the left and then omit that percent sign. So we get .58. Questions so far? (Pause 4 seconds) Here's another example, the one thing that I really want you to remember is that no matter how large the number is, we're only ever going to move the decimal two places to the left. So if I have 258 %, once again my decimal's not included so I put it at the back and I move that decimal two places to the left, so I get 2.58. Now a lot of times this is nice because in popular media, often times they'll use percents. As you get deeper into your academic work, whatever you're going for, in research which you might encounter online or in journals, a lot of times they'll use this notation to represent percents. So instead of saying that 75 % of a population did something, they'll say .75 of the population did something and it's good to be able to translate it back and forth so you can understand it. This is the example that trips people up. You've got 6 % (writes on board 6 seconds). Now we've just followed the rule, then we won't make a mistake; it says move the decimal point two places to the left. The decimal point is at the back of the six. We move it once, twice. Remember that if it's empty, we have to add a zero, so we get .06 (pause 3 seconds). A lot of times people want to say it's this and go (writes .6 on board). What's wrong with this one?

F: That's 60 %.

Erik: Not two places. And it's what?

F: 60 %.

Erik: 60 %, so we don't want to do this [points to .6]. We want to do this [points to .06]. Any questions so far? (Pause 4 seconds) All right. So I've got 43 %; convert that to a decimal (pause 12 seconds). (Erik, Year 1, Basic Math, Observation 1, lines 12–48)

The rest of his lesson unfolded in the same way, with Erik moving along in his presentation, frequently stopping so students could try things at their desks and providing guidance on the procedure for the tasks. Students contributed answers, Erik praised the students and corrected whenever it was necessary. Later in the lesson, he proposed a population activity so students would work with large numbers; students would need to convert from percents

to decimals and practice the use of scientific notation. Unfortunately he brought four-oper-
ation calculators, which could not take the large numbers he was using in the task. The
lesson ended with Erik asking students not to do that problem. He appeared frustrated.

What is the mathematics curriculum generated in this short interaction between
Erik and his students? Ostensibly the content at stake is the conversion between
percentages and decimals. In this lesson students are told that "conversion" involves
moving the decimal point two places to the left and omitting the percent sign. Both
steps in this process are meant to reinforce the direction of the conversion, from
percents to decimals, with decimals presumably being recognized as numbers with
decimal points. This enactment of the lesson does not explicitly discuss the mathematical argument that justifies using the procedure of moving the decimal point two
places to the left or right; Erik only says that this process is equivalent to multiplying by 1/100 or by .01. Erik brings some heuristics for recognizing whether a conversion is right or not ("Remember that if it's empty, we have to add a zero, so we
get .06."). In terms of competencies, he wants students to recognize and develop
their own mathematical power by becoming more confident with doing mathematics; one mechanism he uses to accomplish this is by asking questions students can
answer. He also wants his students to see themselves as doing mathematics for some
more advanced purpose ("in research which you may encounter online"). Although
the population problem was a reasonable activity to ask students to do, managing
their frustration, so they feel competent was more important for Erik. It is reasonable then that he asks students to forget that task.

Erik's own life story drives his views of how students learn, what the best way to
teach is, and what the society values. Like most of his students, he says, he dropped
out of high school. Like them, he survived in odd jobs until he realized that he
needed to learn mathematics in order to become a "sound techie." He started with
arithmetic at the nearby community college and took all the necessary courses. His
trigonometry instructor changed his perception about the importance and beauty of
mathematics because the teacher connected the mathematics with real-life examples
and kept telling them how what they were doing in the course would be so useful for
more advanced mathematics, which would led to better paying jobs. After obtaining
his sound technician certificate, Erik finished his college degree in secondary mathematics education in a night program, enrolled in a master's program in education,
and started teaching, as an adjunct in a community college.

Erik thinks that students learn best by seeing clear explanations of processes;
they gain proficiency by practicing the processes many times. His lessons included
many short segments in which students worked out problems on their own and
called him over to answer individual questions. He praised all contributions and
affirmed the idea of "in-time support," helping students when they needed him. In
terms of pedagogy, Erik worked very hard at selecting examples that would help
develop more complex procedures, interspersing problems that would reinforce the
rules, highlighting special cases (e.g., converting 6 % into a decimal), and asking
questions to maintain student engagement in the presentation. He also foreshad-
owed the importance of mathematics in real life and made statements that brought
the outside world closer, making it appear attainable ("As you get deeper into your

academic work, whatever you're going for"). He knew that his department valued procedural fluency (Kilpatrick et al. 2001) and that this fluency determined whether students were ready for the higher math courses or not.

College Math: Mutually Exclusive Events

Elena smiles broadly and greets her students by name as she passes out a quiz as they come into the room. It is Friday, 9 am, and there are about 15 students in her College Mathematics class, a course required to fulfill the quantitative requirement in many non-science, technology, mathematics or engineering programs. The majority of the students look in their early 20s, although there is a middle-aged man in fatigue clothing sitting by two women in their 40s, one of them on a wheelchair. In the front row there is a woman in her 60s (we learned later in the lesson that she has turned 60 that week) sitting by a young African American girl who is finishing high school. The class resumes 15 minutes later; there are about 30 students in the room now.

Elena: Ok. So you've done your reading and you've looked at the idea of mutually exclusive events. So that's something that you know (writes on board 9 seconds). And it has a pretty simple definition. It cannot happen together (writes on board 7 seconds). So if my experiment is pick one card in the deck (writes on board 6 seconds), pick one card from a standard deck, then the probability, well I don't know that I want to talk about probability yet, I want to talk about the outcome of hearts and seven. Can those happen together?

F: Yes.

Elena: Can you pick a card and have it both a heart and a seven?

Many: Yes.

Elena: So are those mutually exclusive?

F: No.

Elena: No. They're not mutually exclusive (inaudible). What about hearts or spades?

F: Mutually exclusive.

Elena: They are mutually exclusive since they cannot happen together because remember we're only picking one card (pause 5 seconds). So these are mutually exclusive; they cannot happen together (pause 4 seconds). What about three, black?

Many: Yeah.

Elena: Mutually exclusive or not mutually exclusive?

Many: Not mutually exclusive.

Elena: Not because they can happen together. So whenever anything's defined with a negative you have to be a little bit careful, slow down and think about it. It's not mutually exclusive because they can get together [on the board there is a 4 by 2 table; rows are labeled "freshman," "sophomore," "junior," and "senior"; columns are labeled "Candyland" and "Roaring 20s"]. So now I'll explain what that [table] is. That is voting for a homecoming theme at the high school and, believe it or not, Candyland was the theme for the high school's homecoming this year; go figure. I don't know. So those are two themes and they had an election and they voted on which one would be. We don't care which one's the theme; we better add that, right? Figure it out.

(Several students talking at once.)

Elena: So if I pick one person could they be a Candyland supporter and a Roaring 20s supporter?

Several: No.

Elena: They could not because we've clearly defined who's in what set. Could they be a sophomore and a Roaring 20s supporter?

Many: Yes.
Elena: Yes. <u>So which of those is mutually exclusive?</u>
F: Candyland and sophomore.
Elena: <u>Candyland and sophomore are mutually exclusive?</u>
F: No. Roaring 20s and sophomore are mutually exclusive.
Elena: Roaring 20s…
(many at once).
F: Roaring 20s and Candyland.
Elena: Those are mutually exclusive. Ok. <u>So why do we care?</u>
M: We don't.
(Laughter.)
M: For our grades.
Elena: We do care, but I will tell you that what I'm about to explain you've already done so
    don't panic (Elena, Year 1, College Math, Observation 1, lines 2-50).

Elena then launches an introduction of probability and the meaning of formulas for calculating the probability of various events. By the end of the session students were working on conditional probability. With a bubbly personality and extreme confidence, she handles the class swiftly, maintaining students' alertness, making sure that not just the front rows participates, and allowing students break into spontaneous discussions at any time during the lesson. Students in her class speak up freely and joke constantly about how difficult the class is. The class is three hours long and meets only once a week. After 90 minutes, Elena organizes the students in small groups and assigns them the homework. Students are supposed to finish the homework right there with the help of the other students, make sure it is correct, and leave. They will meet again the following week, take a test, and move to the next topic in the course. The course covers 7 different topics on mathematical literacy (e.g., probability and statistics, logic, interest); each topic is dealt with in two weeks, at the end of which there is a test on the topic; there is a quiz in between.

There are two points in Elena's vignette that are important in terms of the curriculum generated in her lesson.

First, she wants to get across an operational definition of mutually exclusive events by drawing students' attention to the familiar context of a card deck. The examples she selected seek to highlight which events "can't happen together" and which can. This is the layman definition that Elena wants students to associate with mutually exclusive and non-mutually exclusive events. By introducing a table, she wants the students to associate a second representation to help students "see" which events are mutually exclusive: the labels of the columns suggest that those are mutually exclusive events; a student can't vote for two themes at the same time. Likewise, the rows are the other mutually exclusive students; a student is a freshman, a sophomore, a junior, or a senior—but not all or some of these at the same time, those "can't happen together." Rather than explicitly connecting the format of the table with the interpretation of the notion, she lets students work out the correct interpretation. The table will be useful later when she talks about conditional probability. In enacting the content of the curriculum in this way, Elena plays an emphasis on helping students make sense of the meaning for the notion of mutually exclusive events.

The second point is that during the interaction, students provide short contributions and answers to Elena's questions (underlined in the excerpt) that move the conversation along. When students propose an incorrect answer, Elena poses the answer back to the class in the form of a question, signaling that the answer is not the expected one. In addition, students only occasionally voice their thinking about a problem or about an answer they found, rather they work out an explanation privately and then publicly provide the correct answer. They explain their reasoning to each other during small groups, but this thinking does not become public.

Elena was a former successful high-school teacher, who had taught a few classes at her nearby community college as an adjunct to increase her income. Over time, she made the transition to the college and has been a full-time faculty in the department for more than 10 years. That she brought her K-12 pedagogical training to the college is evident in the ways in which she organizes the students around collaboration and participation, her focus on practicing procedures and making sense of them, and her attention to grades. For Elena, gaining strategic competence regarding the use of the procedures is important, as she sees these as crucial for success in mathematics. She believes that students need to talk to each other about their thinking in order to be able to learn. At the same time, the responsibility for mathematical justifications and explanations are hers alone. That students only volunteer correct responses publicly seems to be tied to her interest in developing students disposition towards mathematics; her validation of correct response acts as an incentive to boost students' confidence. The grades, which ultimately determine whether the students can continue or not, together with her awareness that the students have difficult lives outside of class, are an important token of exchange in these classes: students accumulate points via participation, quizzes, homework, and exams, and these are all meant to help them build a passing grade.

Trigonometry: Frequency and Phase Shift

Emmett enters his room quietly and sets his book on the front table. He greets the 8:30 am class courteously and starts his 85 minutes long lesson on applications of trigonometric functions. The almost 30 students are also quiet, they sit in rows, more or less close to each other, and appear ready for the class. After solving some homework problems for which students requested help, Emmett states the topic of the lesson, applications of waves, which according to Emmett are electricity, optics and water waves, circuits, and harmonic motion. He defines electric and magnetic fields, using examples from everyday life (rubbing a comb, tightening the belt in his car so the alternator would charge) and the connection between an electric field and an electric current. The students and I are mesmerized by his presentation; he asks occasional questions, that one student in the front row quickly answers. In this lesson he is presenting applications of sine waves, in particular the use of the expression $y = A\sin(Bx + C)$ on p. 156 in their book. Thirty minutes into his presentation he poses the following problem from the homework section in the book:

An alternating current generator produces an electrical current (measured in amperes) that is given by the equation $I = 35 \sin(40\pi t - 10\pi)$ where t is time in seconds. What are the amplitude, period, frequency and phase shift of the current? (Barnett et al. 2006, p. 169)

Emmett: They gave you the equation of current directly; we didn't have to figure it. Ok, first they want the amplitude. The amplitude is the maximum current. How much is the amplitude?

M: 35.

Emmett: 35. It's 35 amps (pause 5 sec). That's the maximum current. Now if you want to graph this, they're asking you to graph it in part b; the height of the graph will go the highest point will be about 35; the lowest point will be minus 35. Next they want the period and the frequency. Ok, the period is 2 pi divided by B, isn't it? [period $= 2\pi/B$, p. 157] 2π over 40π. That's 1/20 which is .05. The period is .05 seconds. So what that means is that the current goes through one entire period every .05 seconds. So every .05 seconds the current is going to repeat itself. It takes on all values from minus 35 to plus 35 every .05 seconds. And then you can start all over. That's what the period means. Frequency, there are two ways to find the frequency. We've got the period [.05] so we can say frequency is 1 over period, which is 1 over .05 and that's 20 hertz. Or we can use the formula for frequency, which is B over 2π. B is 40π; 40π over 2π is 20; that's 20 hertz. And what it means is that the current goes up and down through 20 cycles in one second. A hertz is a cycle per second. 20 times per second up and down; that's what that means (pause 7 sec). We've got one more thing to determine and that's the phase shift. The phase shift, if we have the phase shift, this formula $[40\pi t - 10\pi]$ equals to 0 or by using minus C over B formula. Ok? $40\pi t$ minus 10π equals 0 $[40\pi t - 10\pi = 0]$. Let's (inaudible). We're going to take the 10π to the right side $[40\pi t = 10\pi]$. What's going to happen to the πs on both sides?

Student 1: Cancel.

Emmett: They cancel. So that leaves us with t equals 10 over 40, which is .25. This is phase shift. It has nothing to do with the period. It means the shift is to the right. It means if you look at the graph $\sin(t)$, just t, the graph of $\sin(t)$ will be shifted to the right by .25, by .25 and that's easy to graph; the $\sin(40\pi t - 10\pi)$. Yes?

Student 2: I don't know if I'm right or wrong about this but I noticed inside the parentheses when the sign is like minus, if you just change it to the opposite that's the way it goes on the graph.

Emmett: That's right, yep, you're right. Remember we said that on the first day. When the sign is negative, the shift is to the right and it's positive. When the sign here is positive, the shift is to the left. That always happens (Emmett, Year 2, Trigonometry, Observation 2, lines 203-232).

Emmett continues the lesson in this fashion, writing problems on the board and finding the solutions with some0 input from the students. The students are listening and taking notes. Students ask a few more questions that Emmett answers directly and swiftly. During the break, the students chat quietly and offer me their impressions of his teaching: "he is the best," "he really explains," "he is very patient with us," "he loves his stuff," and "he really wants us to understand."

Emmett chose a problem that would illustrate how mathematical terms have an interpretation in a nonmathematical situation. In terms of curriculum, the lesson seeks to give contextual meaning to various parameters of a "wave" function. Emmett sought to clarify the meaning of the terms (e.g., "20 times/s up and down, that's what that means"). His emphasis during class was on procedure execution rather than on other possible mathematical competencies such as giving a mathematical justification for why the procedures work (e.g., explanation for shift to the right instead of the left); making connections to other solutions, representations, or problems (e.g., why are the two procedures to find frequency equivalent?); analyzing the situation to respond questions on their own (e.g., what procedures need to be done?); evaluating a situation (e.g., why is one procedure better than the other?); or producing new knowledge (e.g., proposing a situation in which the frequency is half of the one given here).

Emmett immigrated to the United States over 30 years ago; he has a bachelor's degree in engineering from his home country and a PhD in physics from an university in the United States. He has been teaching at the college level for 20 years and was a trainer for an engineering firm for two. Emmett consistently brought his background in engineering and physics to contextualize notions he was explaining. During interviews, Emmett indicated that he saw students' questions as symptoms of a weak preparation: students in remedial courses need to ask more questions because they have more difficulties with the content; students in more advanced courses understand better and therefore do not need to ask that many questions in class. His background in physics and engineering, though, made evident what he believed should be valued in the courses, applications, engineering, and sciences. Like Erik and Elena, Emmett emphasized procedures; differently than them, he did not encourage students' questions or used class time to solve problems. He assumed students learned best by seeing applications and making sense of the terminology in context; his role was to be clear and very explicit about the contextualization. Ultimately, his conviction that to have a job students needed to see mathematics being used, drove his emphasis on applications in mathematics.

Discussion

La historia, la sociedad, la política y la cultura trabajan juntas para asegurar una gran estabilidad de los currículos a través del tiempo y a través de las escuelas en un país, sin considerar el flujo de los mandatos que provienen de las oficinas centrales [History, society, politics, and culture all work together to ensure great curriculum stability across time and through all schools in a country without considering the mandates coming from central offices]. (Kilpatrick 1998, pp. 105, trans. by the author)

When I examine these vignettes through the lens of the generated curriculum, I wonder about the prospects for a reform that could be guided via curriculum in the community college context. This quote is fundamental because it acknowledges that the reform does not simply happens by a mandate given by a central office or an institution. This quote reminds us that curriculum is stable. These vignettes, drawn from the community college context, illustrate that the stability is a consequence of the historical, societal, political, and cultural conditions that surrounds the generation of the curriculum.

The community college context in which Erik, Elena, and Emmett teach helps us see the notion of stability more clearly. These instructors enact a vision of curriculum in their classroom that is shaped by their own history, the needs of our society, our current political environment, and the U.S. culture. Indeed, during interviews, these teachers indicated being inspired by their own personal experiences with great teachers, who provided clear explanations, were patient, showed real-world applications or connections, used multiple examples, did not embarrass them in the classroom, and made the material look easy. In their own experience, these teachers opened doors for them to continue on studying more mathematics and attain a successful position in life. Now, as instructors in this institution, they work towards emulating their vision of what great teachers did for and with them, with the conviction that their students can repeat that history of success, within the parameters offered by their college.

Unlike 4-year institutions, community colleges have been described as institutions that can respond easily to changes in society (Alfred et al. 2009; Cohen and Brawer 2008; Jacobs 2011). They can provide training in many professions that don't require a four-year degree and adapt quickly to generate programs that can fulfill the demands of the community. Indirectly, mathematics departments have benefited by these changes, as more programs do require basic literacy in mathematics, specifically some algebra and data management and analysis. There are pressures for having more students take and pass these classes, but not necessarily for a distinct type of mathematics to be taught. Mathematics department have responded to the increasing number of students enrolled by offering more sections of mathematics courses taught largely by adjunct instructors, whom tend to have little contact with their full-time colleagues and who, are for the most part, left on their own to teach.

The current political discourse calls for more students to attain a college degree; a short-term translation of this vision has resulted in increased attention to passing rates in basic courses as a predictor of success (staying in a program and finishing it) and in discussions about possibly turning away students who do not place in their basic courses (i.e., becoming more selective). This increased attention to passing rates and selectivity is another expression of a credentialing system (Labaree 1997), resulting in a process that privileges moving through the courses demonstrating competency on very limited terms over making sure that students learn content in a permanent way.

The community college has been labeled the last truly democratic educational institution in he United States and the only one left called to fulfill the dream of educational opportunity for all (Bailey and Morest 2006; Morest 2006). Community colleges embody an important value in he U.S. culture, namely, that all people deserve an opportunity for social mobility. Community colleges are enrolling more minority, low-income, English as a second language, first-generation college, and disabled students than 4-year colleges or universities (Goldrick-Rab 2007). Across the United States, the open admissions policies make it possible for many students to pursue studies that are inaccessible to them in other countries and in other American postsecondary institutions. But, one can argue, this very same notion of always having an opportunity impacts how students and teachers perceive their role in schooling and influences their perception of the curriculum. In principle, all topics could be revisited later, with a different emphasis and for a different purpose. In principle, students and teachers will encounter each other teaching similar ideas giving all a new opportunity to both master and to teach the ideas in the curriculum. A stable curriculum gives students and teachers the possibility of fulfilling each other's goals.

The curriculum generated through these vignettes emphasizes what some have called "inert," "instrumental," or "surface" knowledge (Erlwanger 1973; Novak 1977, 1998; Skemp 1971; Trigwell and Ashwin 2006; Trigwell and Prosser 2004) a type of knowledge that is important but in general insufficient to ensure adequate student learning (Anderson et al. 2001; Bransford et al. 1999; Hiebert and Carpenter 1992; Sfard 1991). This curriculum attends primarily to procedural proficiency, with limited attention to conceptual understanding or interest in promoting students' productive disposition by using mathematical reasoning. Likewise, as the vignettes illustrate, strategic competence and adaptive reasoning are rarely discussed in these lessons. Thus, proposals, such as those in the *Adding It Up* report (Kilpatrick et al. 2001), that have clearly described a vision for mathematics proficiency at all levels appears foreign in these classrooms.

A curriculum reform that expects generating a different curriculum will require engaging teachers, students, and the mathematical content in ways that run against these historical, social, political, and cultural forces; these forces work together to have a system that facilitates students' progress towards degree attainment at the expense of learning authentic mathematics (see Fig. 7.2). In terms of the definition of curriculum, these forces act against mandates dictated by central offices or organizations that seek to reduce number of courses students take in their programs (see, e.g., AMATYC's *Beyond Crossroads,* AMATYC's *New Life initiative,* http://www.devmathrevival.net/?page_id=8, or the curriculum initiatives by the Carnegie Foundation for the Advancement of Teaching, http://www.carnegiefoundation.org/developmental-math).

What we learn from this rumination on Jeremy's quotes is that unless we closely examine how the curriculum is generated in the daily interactions with students,

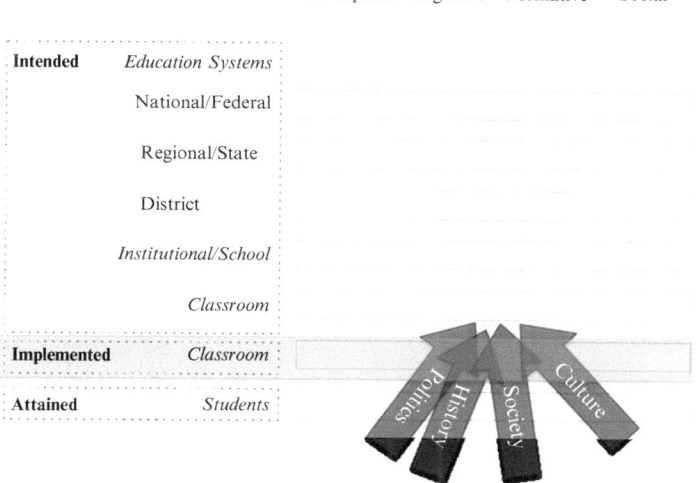

Fig. 7.2 Forces that help maintain a stable generated curriculum

very little will change for our students. We need to heed the lessons from history that recommends that we take seriously the wider social, political, cultural, and economic contexts in which curriculum is created, generated, and experienced. As Jeremy suggests, not doing so may indeed result in yet another failure of our efforts to bring about reform in mathematics education in this setting.

Coda

I teach a research in mathematics curriculum course in our mathematics education program. Like Jeremy, I start the class by asking my students to define curriculum. Like in my first curriculum class I get a wide range of responses that assume an external body of content, "pre-existing somewhere 'out there' in need of being transmuted into an elementary form that teachers and students can encounter it." Throughout the course, the students and I work towards understanding how in each setting it is necessary to understand the many forces that impinge on the implementation of the curriculum, which is what becomes students' actual experiences with it. We reach the end of the semester with some frustration but with better understanding of the meaning of curriculum and its uses for reform.

References

Alfred, R., Shults, C., Jaquette, O., & Strickland, S. (2009). *Community colleges on the horizon: Challenge, choice, or abundance*. Lanham: Rowman & Littelfield.

Anderson, L. W., Krathwohl, D. R., Airasian, P. W., Cruikshank, K. A., Mayer, R. E., Pintrich, P. R., & Wittrock, M. C. (Eds.). (2001). *A taxonomy for learning, teaching, and assessing*. New York: Longman.

Attewell, P., Lavin, D., Domina, T., & Levey, T. (2006). New evidence on college remediation. *J High Educ, 77*, 886–924.

Bahr, P. R. (2008). Does mathematics remediation work? A comparative analysis of academic attainment among community college students. *Res High Educ, 49*, 420–450.

Bailey, T. R., & Morest, V. S. (2006). *Defending the community college equity agenda*. Baltimore: Johns Hopkins University Press.

Barnett, R. A., Ziegler, M. R., & Byleen, K. E. (2006). *Analytic trigonometry with applications* (9th ed.). Hoboken: Wiley.

Biehler, R., Scholz, R. W., Strasser, R., & Winkelman, B. (Eds.). (1994). *Didactics of mathematics as a scientific discipline* (Vol. 13). Dordrecht: Kluwer.

Blair, R., Kirkman, E. E., & Maxwell, J. W. (2013). *Statistical abstract of undergraduate programs in the mathematical sciences in the United States. Fall 2010 CBMS Survey*. Washington, DC: American Mathematical Society.

Bransford, J. D., Brown, A. L., & Cocking, R. R. (1999). *How people learn: Brain, mind, experience, and school*. Washington, DC: National Academy Press.

Clark, B. R. (1960). The "cooling out" function in higher education. *Am J Sociol, 65*, 569–576.

Cohen, A. M., & Brawer, F. B. (2008). *The American community college* (5th ed.). San Francisco: Jossey-Bass.

Dowd, A., Bensimon, E. M., Gabbard, G., Singleton, S., Macias, E., Dee, J. R., ... Giles, D. (2006). *Transfer access to elite colleges and universities in the United States: Threading the needle of the American dream*. Jack Kent Cooke Foundation, Lumina Foundation, Nellie Mae Education Foundation, University of Massachusetts, Boston, MA.

Erlwanger, S. H. (1973). Benny's conceptions of rules and answers in IPI mathematics. *J Math Behav, 1*(2), 7–26.

Goldrick-Rab, S. (2007). *Promoting academic momentum at community colleges: Challenges and opportunities. CCRC Working Paper No. 5*. Madison: University of Wisconsin.

Gómez, P., & Fernández, F. (1997). Graphing calculator use in precalculus and achievement in calculus. In E. Pehkonen (Ed.), *Psychology of mathematics education* (Vol. 3, pp. 1–8). Lahti: University of Helsinki.

Gómez, P. & Mesa, V. (Eds.). (1995). *Situaciones problematicas en pre-calculo* [Problematic situations in pre-calculus]. México y Bogotá: Editorial Iberoamerica & "una empresa docente".

Gómez, P, Carulla, C, Castro, M, Fernández, F, Gómez, C, Mesa, V. ... Valero, P. (Eds.). (1995). *Aportes de "una empresa docente" a la IX CIAEM* [Contributions of 'una empresa docente' to the IX Inter-American Conference in Mathematics Education]. Bogotá: una empresa docente.

Grubb, N. W. (1999). *Honored but invisible: An inside look at teaching in community colleges*. New York: Routledge.

Hiebert, J., & Carpenter, T. (1992). Learning and teaching with understanding. In D. Grouws (Ed.), *Handbook of research on mathematics teaching and learning* (pp. 65–97). New York: Macmillan.

Jacobs, J. (2011, February 7). Hard lessons learned from the economic recession. *Community College Times*.

Kilpatrick, J. (1997). Reviews: Didactics of mathematics as a scientific discipline. *J Curriculum Stud, 29*(6), 735–738.

Kilpatrick, J. (1998). Cambio local y global del currículo [Curriculum change locally and globally]. *Rev EMA, 3*(2), 99–111.

Kilpatrick, J., Swafford, J., & Findell, B. (2001). *Adding it up: Helping children learn mathematics*. Washington, DC: National Academy Press.

Labaree, F. D. (1997). *How to succeed in school without really learning: The credentials race in American education*. New Haven/London: Yale University Press.

Lutzer, D. J., Rodi, S. B., Kirkman, E. E., & Maxwell, J. W. (2007). *Statistical abstract of undergraduate programs in the mathematical sciences in the United States: Fall 2005 CBMS Survey*. Washington, DC: American Mathematical Society.

Melguizo, T., Hagedorn, L. S., & Cypers, S. (2008). Remedial/developmental education and the cost of community college transfer: A Los Angeles county sample. *Rev High Educ, 31*, 401–431.

Mesa, V., & Gomez, P. (1996). Graphing *calculators and precalculus: An* exploration of some aspects of students' understanding. In L. Puig & A. Gutierrez (Eds.), *Proceedings of the 20th conference of the International Group for the Psychology of Mathematics Education* (Vol. 3, pp. 391–398). Valencia: Universitat de Valencia.

Mesa, V., Gómez, P., & Cheah, U. (2013). Effects of international studies of student achievement on mathematics teaching and learning. In M. A. Clements, A. Bishop, C. Keitel, J. Kilpatrick, & F. Leung (Eds.), *Third international handbook of mathematics education* (pp. 861–900). New York: Springer.

Morest, V. S. (2006). The attempt to balance multiple missions. In T. R. Bailey & V. S. Morest (Eds.), *Defending the community college equity agenda* (pp. 28–50). Baltimore: Johns Hopkins University Press.

National Council of Teachers of Mathematics. (1989). *Curriculum and evaluation standards for school mathematics*. Reston: National Council of Teachers of Mathematics.

National Council of Teachers of Mathematics. (1991). *Estándares curriculares y de evaluacion para la educación matematica*. Sevilla: Sociedad Andaluza de Educación Matemática.

Novak, J. D. (1977). *A theory of education*. Ithaca: Cornell University Press.

Novak, J. D. (1998). *Learning, creating, and using knowledge*. Mahwah: Erlbaum.

Perry, P., Gomez, P., & Valero, P. (1996a). The teaching of mathematics from within the school: Teachers and principals as researchers. In L. Puig & A. Gutiérrez (Eds.), *Proceedings of the 20th conference of the International Group of the Psychology of Mathematics Education* (Vol. 4, pp. 123–130). Valencia: Universidad de Valencia.

Perry, P., Mesa, V., Fernández, F., & Gómez, P. (1996). *Mathematicas, azar, sociedad: Conceptos básicos de estadística* [Mathematics, chance, and society: Basic concepts in statistics]. Mexico & Bogotá: Editorial Iberoamérica & una empresa docente.

Rodi, S. B. (2007). *Snapshot of mathematics programs at two-year colleges in the U.S.* Available at http://www.amatyc.org. Austin: Austin Community College.

Sfard, A. (1991). On the dual nature of mathematical conceptions: Reflections on processes and objects on different sides of the same coin. *Educ Stud Math, 22*, 1–36.

Skemp, R. (1971). *The psychology of learning mathematics*. Harmondsworth: Penguin.

Travers, K. J., & Westbury, I. (Eds.). (1989). *The IEA Study of mathematics I: Analysis of mathematics curricula*. Oxford: Pergamon Press.

Trigwell, K., & Ashwin, P. (2006). An exploratory study of situated conceptions of learning and learning environments. *High Educ, 51*, 243–258.

Trigwell, K., & Prosser, M. (2004). Development and use of the Approaches to Teaching Inventory. *Educ Psychol Rev, 16*, 409–424.

Waycaster, P. (2001). Factors impacting success in community college developmental mathematics courses and subsequent courses. *Commun Coll J Res Pract, 25*, 403–416.

Part III
The Interaction of Theory, Practice and Politics in Mathematics Education

Chapter 8
Mathematics, Knowledge and Political Power

Christine Keitel-Kreidt

Abstract In this chapter, I trace the historical evolution of conceptions of the social impact of mathematics, giving special attention to the influence of modern information and communication technologies. I explore links between this development and the corresponding evolving conceptions of mathematical literacy and mathematics education for all.

Keywords Mathematics • Mathematical literacy • Information technology

Introductory Remarks Jeremy Kilpatrick is one of the rare colleagues who supported and still support strongly political essays and debates about most important themes and debates in mathematics education. His support was a very important feedback for my perceptions of mathematics education as a global political enterprise, which I dared to follow up quite some years and still do. All the years, it was and still is very helpful to get encouragement by a colleague with such great visions. The following essay is designed to honour Jeremy Kilpatrick.

Mathematics is perceived today not only as one of the most powerful social means for creating new technologies but also one of the most powerful social means for planning, optimising, steering, representing and communicating social affairs created by mankind. By the development of modern information and communication technologies (ICT) based on mathematics, this social impact of mathematics came to full power: Mathematics is now universally used in all domains of society, and there is nearly no political decision-making process, in which mathematics is not used as the rational argument and the objective base replacing political judgements and power relations.

However, for the ordinary citizen, it becomes increasingly difficult and sometimes impossible to follow these developments of mathematics, mathematical

C. Keitel-Kreidt (✉)
Fachbereich Erziehungswissenschaft und Psychologie,
Freie Universität Berlin, Berlin, Germany
e-mail: keitel@zedat.fu-berlin.de

© Springer International Publishing Switzerland 2015
E. Silver, C. Keitel-Kreidt (eds.), *Pursuing Excellence in Mathematics
Education*, Mathematics Education Library, DOI 10.1007/978-3-319-11952-6_8

applications and ICT and to evaluate their social use appropriately, because specialisation and segmentation of mathematical applications often are extremely hard to understand. The principal insight into their necessity and a basic acknowledgement of their importance in general are often confronted by a complete lack of knowledge of concrete examples of their impact. Competencies to evaluate mathematical applications and ICT, and the possible usefulness or its problematic effects, however, now are a necessary precondition for the political executive and the democratic participation of citizens. The new challenge is to determine what kind of knowledge about social knowledge in a mathematised society is needed and how to gain the necessary constituents.

Studies in the history and philosophy of sciences show that *changes of forms of symbolising and processing of information* usually have mutual *effects to social organisation* and lead to *new structures of scientific knowledge* (Damerow and Lefévre, [Reckoning coins, experiment and language. Historical case studies on the beginning of the exact sciences]. 1981; Renn, [Science as orientation for life – A story of success?]. 2002).

Together with the change of structures of knowledge, characteristic *styles of thinking and in particular worldviews* are developed, which usually cause also most fundamental changes in the process of political decision-making on general social goals and means and measures to pursue them, in particular with effects for the allocation of resources in a society.

Historical Accounts of Mathematics and Political Power

Mathematics as a Distinctive Tool for Problem-Solving in Social Practices and Means of Social Power

Historical studies show that since the beginning of social organisation, social knowledge of exposing, exchanging, storing and controlling information in either ritualised or symbolised (formalised) way was needed, therefore developed and used, in particular information that is closely related to production, distribution and exchange of goods and organisation of labour (Davis & Hersh, 1986).

This is assumed as one of the origins of mathematics: Early concepts of number and number operations and concepts of time and space have been invented as means for governing and administration in response to social needs. Control of these social practices and the transmission of the necessary knowledge to the responsible agents were mostly secured by direct participation in social activities and direct oral communication among the members. Ritualised procedures of storing and using information have been developed since Neolithic revolution, during the transition to agriculture and permanent living sites, which, e.g. demanded planning the cycles of the year.

The urban revolution and the existence of stratified societies with a strong division of labour induced symbolical storage and control of social practices by information systems based on mathematics, which were bound to domain-specific systems of symbols with conventional meaning. The earliest documents available are the clay tablets from Mesopotamia (Uruk 3000 BC), in which mathematics appears as necessary and useful tool for solving problems of agriculture and economic administration – 'bookkeeping' of production and distribution of goods in a highly hierarchically structured slavery-based society (Nissen et al. 1990). We witness mathematics of that time as employed as a technique and a useful and necessary tool. The scribe, who disposed of the appropriate knowledge to handle this tool, becomes an important man. More generally speaking, the governing class or group disposes of mathematics as an additional instrument of securing and extending its *power and authority*.

Mathematics as Theoretical System and a Base of a Universal Political Worldview

A new and eventually most consequential perspective of mathematics emerged in Ancient Greece: mathematics (more correctly: geometry) as a theoretical system, as a philosophy, as the queen of sciences and as a universal divine mental force for mankind. Greek societies were differentiated into two classes with two distinct social practices: the non-Greeks or slaves for all practical and technical manual labour necessary for the maintenance and practical life of the society and the Greek citizens for warfare, physical sport activities as leisure and spiritual activities in politics, philosophy, rhetoric and the other of the seven 'liberal arts', mathematics among them. For the Greek, therefore, mathematics was detached from the needs of managing ordinary daily life as from the necessity of gaining their living. Instead, by scientific search for fundamental, clearly hierarchically ordered bases, creating connections and elements of a systemic characterisation of existing formal mathematical problem-solving techniques and devices, independent of any specific practical intention, they reformulated mathematics as a scientific system and philosophy, a (Platonic) ideal theory to be further discovered and constructed by human theoretical thinking and reasoning, not by doing or solving practical problems. This distinction between *mathematics as the queen*, as a science of formal systems by introducing a structure and defining mathematical thinking as logical reasoning from axioms to concepts and theorems to proofs, in opposition to a view of mathematics as a simple technique or problem-solving tool, which is only and simply used, *mathematics as the servant*, is ascribed to Greek scholars (Snell 1948). Being able to think mathematically was a sign of those who have political power. By viewing mathematics as the structure underlying the construction of the cosmos and *number* as the basis of the universe and emphasising a hermetical character of the mathematical community, the ground was laid for the high esteem of mathematics as a segregation means of political and social power by the Christian church in the middle ages.

Mathematics as a Human Force for General Scientific and Social Development

Over the centuries, the traces of structuring the world by human rational activity became more numerous, appropriate dealing with it, more imposing. There were several fields in which the mathematisation of the real world and of social life advanced more remarkably, among these notably architecture, military development in both, fortification and armament, mining industry, milling and water regulation, surveying and, before all, manufacturing and trade. The extension of trade from local business to far distances exchange prompted the emergence of banking, and for the functioning of this, an unambiguous form of clear and universal regulation was needed: the system of bookkeeping was invented. *This was the first consistent, comprehensive mathematical structuring of a whole field of a social practice* (Damerow et al. 1974).

All these endeavours culminated in the period of the European Renaissance with a unique confluence of a wide range of contributions: inventions, discoveries and human genius. The rediscovery of Greek culture incited a revolutionary change of perceptions, and their secularisation, the idea of man as an autonomous individual and a merging of all of his various capacities and powers in this one notion of the individual genius. A prototype of this *new man* is Leonardo da Vinci, painter, architect, mathematician, engineer, inventor, scientist, writer, cartographer, etc. A key achievement in renaissance mathematics is linear perspective, and interestingly, it is in this point that renaissance mathematics and art converge. It is not surprising if we identify Leonardo's activity in both art and mathematics as visual research. Mathematics is thus seen as the queen *and* the servant of sciences, as a practical *and* theoretical tool, as artful theory and general philosophy and a base for the development of technology and natural sciences, with a worldview to "discover the world" to the benefit of all citizens and "to tame or dominate nature" (cf. Kline, 1985).

Mathematics as Rationality and Common Sense

By the development of universities, (mathematical) knowledge emancipated from clerical purposes. Renaissance interest in antique culture also contributed to rediscover and re-edit classical texts and old knowledge, and book printing made them available to a wider public. The technology of unlimited reproducibility of knowledge by the revolution of media development by book-printing machines – a revolution like the more recent one by modern radio and television broadcasting and ICT – enabled and demanded to decide about standards and canonical representations of knowledge. The sciences emancipated from religious and philosophical restrictions, as mathematics from religious and philosophical bonds. The abundance of knowledge became itself a subject of analytical reflection and of new philosophical approaches.

The idea of the '*rational man*' developed: When developing algebra as a general method for mathematics and rational thinking, Descartes believed that mathematics itself could become so 'easy' or easily understandable, accessible and acceptable by all people that it can be considered as part of '*common sense*', 'le bon sens pour tout le monde' (Davis & Hersh, 1986). Leibniz, one of the inventors of calculus, shared this perspective of the rational mankind: In evaluating the discovery of calculus, he believed that rational discourse and strict mathematical reasoning have become unlimited and would solve all social and political problems in the world. His call '*Calculemus (let us compute*!)' encourages those engaged in a dispute to turn it into computing, and stated that whenever and wherever a dispute arose, calculation should solve it, and finally save the whole world from controversies, from hostile actions and even from war! The application of a mathematical, rational argumentation and calculation was considered as the universal remedy for any personal or social problem, as it solved problems in a way understandable and acceptable for everybody and accessible for rational proceeding. Mathematics and thinking mathematically were considered as the fundament of a 'sane' mind building, as a general reasoning competency, the facilitator and creator of rationality and the *rational mankind*.

At the same time, mathematics had become more and more a necessary tool for the development of *scientific knowledge* and *craft knowledge*, the *professional knowledge for practitioners*. The increasing importance of trade and commerce demanded extensive computation skills in trade and commercial and banking companies, but also in manufactures, and quality control of production and distribution necessitated new mathematical tools. The availability of Arab-Indian mathematics and their connotation system allowed for *written computation* with cyphers, decimal fractions and *formal solutions for practical problems* of craft, trade and commerce in terms of *calculation rules*, and appropriate schooling was demanded and propagated. Computation schools served as a secular complement and a necessary element in various kinds of vocational or professional training in practical mathematics.

Social Needs for Mathematics Education

Professionalisation and Specialisation of Knowledge

The achievements of the fifteenth to eighteenth century entail an explosion of trades, crafts, manufacturing and industrial activities with an impressive diversity, ingenuity and craftsmanship (mostly mechanical) developed and required in numerous professions. The ability of a greater part of population to appropriately deal with fundamental systems of symbols like writing and calculating becomes a *condition for the functioning of societies*: Elementary (mathematics) education and training is established as reaction to social demands and needs, either prior or during vocational training and various professional practices.

Parallel to upcoming educational institutions and in concert with them, mass production for unlimited reproduction of knowledge *enables* and *needs standardisation* and *canonical bodies and representations of knowledge*. A reflection and restructuring of existing knowledge on a higher level is asked for: Meta-knowledge has to be developed that offers standards of knowledge and their canonical representations for *educational purposes*; at the same time, meta-knowledge as orienting knowledge becomes an immanent condition for developing *new systems of knowledge*, in particular for sciences like mathematics that are perceived to a greater part as independent of immediate practical purposes.

In the nineteenth century, the competition between the bigger European states, inspired by a strong and fateful ideology of national superiority and ambition, draws attention to *'knowledge as power'*, making school education a central interest of governments. Industrialisation is accompanied by an increasing autonomy of systems of scientific and practical knowledge. *To be a mathematician*, somebody who does mathematics and nothing else, is a new profession, conceived as autonomous, without immediate practical use in other domains, mathematicians as scholar at university or a high-level school teacher. As in sciences, specialisation and professionalisation of experts become requirements in all branches of economy, as in social services and administration. Constructing and creating new knowledge become *a precondition for the material reproduction of society, not consequence*. Specialisation is a condition for creating new knowledge but at the same time bears the risk of disintegrating more comprising systems of knowledge and making integration in a wider context difficult. *Partial knowledge* must be generalised and incorporated on a level of *meta-knowledge*.

(Mathematics) Education as a Public Task

In the nineteenth century in many countries, public- and state-controlled two partite school systems are created: higher education as *mind forming for an elite*, elementary education to *transmit skills and working behaviour for the majority*, the future working class.

Humboldt's notion of 'Bildung' comprises learning as universal as ever possible with strong emphasis on humanities: philosophy, history, literature, art, music, but also with an emphasis on mathematics and sciences. The ideal was the completely cultivated, best educated human being, and 'Bildung' was not a process ending at the end of one's studies, but just the base laid in the youth to be enlarged and enriched during the whole life, 'Bildung' as *an attitude and a path as much as an accomplishment*. And that was to be conveyed by means of a public education, at secondary schools and at universities. *Mathematics* becomes subject in higher education institution *for the elite and governing class* because of its formal educational qualities, e.g. educating the mind independent of a direct utilitarian perspective and fostering general attitudes *to support the scientific and science-driven technological development*.

In the elementary or general school for future workers and farmers, only arithmetic teaching in a utilitarian sense is offered: to secure the necessary skills for the labour force and to secure acceptance of formal rules and formal procedures set up by others. Mathematics education for the few is strictly separated from the skill training for the majority; this corresponds to a separation of mathematics education as an art and science in contrast to mathematics education as a technique, *scientific knowledge and conceptual thinking versus technical, algorithmic, machine-like acting*.

Mathematics Education for All

In the nineteenth and twentieth century, mathematics became the driving force for almost all scientific and technological developments: mathematical and scientific models and their transformation into technology impact not only on natural and social sciences and economics but also on all activities in the social, professional and daily life. This impact increased rapidly by the development of the new information and communication technologies (ICT) based on mathematics, which radically changed the *social organisation of labour* and our *perceptions of knowledge or technique* to an extent that is not yet fully explored.

On the one side, mathematics as a human activity in a social environment is determined by social structures; hence, it is not interest-free or politically neutral. On the other side, the continuous application of mathematical models, viewed as universal problem-solving procedures, provides not only descriptions and predictions of social actions but also prescriptions: The increasing social use of mathematics makes mathematical methods and ways of argumentation to *quasi-natural social rules and constraints* and creates a mathematised *social order* effective in social organisations and hierarchical institutions like bureaucracy, administration, management of production and distribution, institutions of law and military, etc. *Social and political decisions are turned into facts, constraints of prescriptions that individual and collective human behaviour has to follow.*

New perspectives of the social role of (mathematical) knowledge and general education were developed that gained political acceptance and support: 'Mathematics Education for All' and 'Mathematical Literacy'. The concepts were differently substantiated and received different interpretations and supporters: The New Math movement had started to introduce mathematics for all by a formally unified, universally applicable body of theoretical knowledge of modern mathematics exposed to all but had to be revisited and discarded as a solution. Intensive work in curriculum development created a wide range of different and more and more comprehensive approaches combining new research results in related disciplines like psychology, sociology and education and developed this vision further (Howson et al. 1981; Sierpinska and Kilpatrick 1998). A variety of conceptions promise to describe the socially necessary knowledge in a more substantiated form and to integrate scientific mathematical practices and common vocational or professional practices and their craft knowledge, or conceptual and procedural knowledge, or mathematical modelling and application.

A new and most radical development within and outside of mathematics as a discipline was caused by ICT, by the invention of electronic media and by the new possibility of data processing. The immediate consequence, based on the integration of human-mental and sensory-information processing techniques within machines, is the creation of *technologies which take over human information processes and independently determine social organisation.* This new development is called *globalisation of knowledge*: the technological integration of new representation forms and the distribution of knowledge in a global net of knowledge represent the greatest challenge for a restructuring of political power and decision-making processes about the way in which information is gained and used, available to anybody everywhere with access to the Internet.

Information and communication technologies are the fundament for communication which is an essential aspect of *globalisation*: access to and exchange of information and knowledge from anywhere in the world, quickly and cheaply. On the one hand, that leads to a general acknowledgement of cultural diversity, but on the other hand also to universalisation and domination by certain languages and cultural positions – e.g. the English language and Euro-American or Western belief systems, encompassing a variety of knowledge traditions and knowledge systems.

Changing Mathematical Sciences and Its Applications by Information Technologies

The social role of mathematics and its social impact has dramatically changed by the development of modern information technologies based on mathematics. Mathematics is ascribed a new utility value, which has never before been as strongly indubitable as it is now. Illustrating examples for new technological and most effective applications of mathematical methods are numerous, e.g. computer-based simulations are applied in most different areas like modelling of climate changes, crash tests and chemical reaction kinetics by building process-oriented technical machines and dynamical system models in macroeconomics and biology.

Software packages allow the most complex calculation processes for many applications in forms of black boxes, like statistical processes in quality control, research on market and products, risk theories for portefeuille management in assurance companies, computer-based algebra systems and software for modelling in sciences and engineering. Mathematics as the basis of many technologies is effective although only invisibly, i.e. as theoretical base of formal language in informatics and as fundament of coding algorithms for industrial robots or in the daily used scanners, mobile phones, cash corners or electronic cashiers. New technologies in return have feedback with great impact on mathematics as a discipline itself. Besides traditionally applied mathematics, new directions combined applied sciences with *experimental procedures* like techno-mathematics, industrial mathematics and theory of algorithms.

New procedures in some application areas are celebrated not only as new means to ends or refined methodological repertoire but furthermore as a *new paradigm*: In contrast to classical applied mathematics, which was oriented towards and restricted to the representation of mathematical structures of a reality existing completely independent of any subjective intention, new forms of applications do not hide the fact that interests and intentions always guide the construction of a model, as well as specific goals and convictions (Davis, 1989). The theoretical poverty of such models is interpreted as advantage, as no comprehensive theories of the object have to be presupposed; by some, the new paradigm is celebrated as humanisation of modelling.

Mathematics and information technology not only provide descriptions and explanations of existing reality, but they also *create new reality*: As a basis of social technologies like arithmetical models for election modes, taxation models, calculation of interests and investment, calculation of costs and pensions, etc., mathematical models are transformed into reality and establish and institutionalise a new kind of reality. This process can be reconstructed and analysed as the development of implicit mathematics: Patterns of social acting and formal structures are transformed via formal languages into algorithms or mathematical models which can be rectified and objectified as social technologies (Keitel et al. 1993; Skovsmose 1994).

In models of macroeconomy, translations of an ideology into mathematical concepts even can be identified, which by enrichment with subtle economical terminology and by internal consistency of the mathematical representation suggest not only progress but existence as a natural law. In such a way, mathematisations also can be established as *unconscious cultural forms and rites* and as a kind of language that creates a milieu for thoughts, which further creates *unquestioned constraints and restrictions of consciousness.*

Such results of applications of mathematics are often encountered in communication situations mainly shaped by conflicting interests where they serve to *justify opinions* and to *stabilise attitudes*. Graphical representations of information, e.g. are excellently structured and provide sufficient overview and relative universality of readability but are also appropriate means for accentuation guiding the perception into wrong directions. In such communication processes, the possibilities for interaction between interpreters are usually restricted. Even neglecting the fact that credibility is often depending on the prestige of the participants, the prestige of mathematics as such often serves to suggest objectivity and objective goals and intentions. Thus, the *regulation and democratic control* of actual and future research, development and application processes of mathematics and mathematics education demand a specific competence and knowledge as a basis of decision-making on the side of the *politicians* and new knowledge for evaluation and democratic control on the side of the *citizens*.

Mathematical Literacy for Critical Citizenship in a Democratic Society

The *pervasiveness of economic thinking and interests* have successively created so high a pressure of economic orientation that educational aims and the subject matter are marginalised unless they prove justification in terms of economic interest (Woodrow 2003).

New notions like '*Mathematical Proficiency, or Competency or Literacy*', '*Educational Standards*' and '*Benchmark*' are expressions of such economic interests. They are a major concern of politicians but also a pressure for educational researchers and practitioners. They are the key issues in the recent political debates and disputes about mathematics education, which broadened after the release of international comparative studies like TIMSS and PISA and their ranking of test results. Proclaiming that the PISA tests are based on 'definitions of mathematical literacy' that are underpinned by fundamental and widely accepted educational research results, and that it is absolutely unproblematic to test such kind of competencies or proficiencies on a global scale to rank countries' performances, produced strange and urgent political measures to be taken in some of the countries that did not perform well, called for by the alarmed public and the medias.

The *Programme for International Student Assessment* (*PISA*) claims for its test of *Mathematical Literacy* that those competencies of young adolescents are measured, which enable them to participate in democratic decision-making processes: '*Mathematical Literacy is the capacity to identify, to understand and to engage in mathematics and make well-founded judgements about the role that mathematics plays, as needed for an individual's current and future life, occupational life, social life with peers and relatives, and life as a constructive, concerned and reflective citizen*' (OECD 2000, 50).

Results of tests like PISA are used as reference and base for decisions in educational policy, in particular in the case when they show that only a small part of the tested students or adults have reached a higher level of competencies in an international comparison. However, each attempt to define Mathematical Literacy is confronted with the problem that this cannot be done exclusively in terms of mathematical knowledge: To understand of mathematised contexts or mathematical applications and to competently use mathematics in contexts goes beyond mathematical knowledge. A first research study to explore such cross-curricular competencies by investigating the ways how mathematics is used in a social–political practice had unexpected and surprising results (Damerow et al. 1974). Mathematical Literacy must be understood as functional in relation to pedagogical postulates. But by reducing the concept of Mathematical Literacy to the descriptions of the process of its measurement cannot be justified, while conclusions of these comparisons mostly are formulated in terms of daily language or connected with highly demanding and complex meanings and connotations of the concepts.

Conflicting conceptions of Mathematical Literacy are numerous, although the conflict is not always recognised: Jablonka (2003) analyses what research on Mathematical Literacy can do and what *not* by investigating different perspectives on Mathematical Literacy. She shows that these perspectives always considerably vary with the values and rationales of the stakeholders who promote them. The central argument underlying each of her investigations is that it is not possible to promote a conception of Mathematical Literacy without at the same time – implicitly or explicitly – promoting a particular social practice of mathematics: be it the practice of mathematicians, of scientists, of economists, of professional practices outside science and mathematics, etc. She argues that Mathematical Literacy focussing on citizenship in particular refers to the possibility or need of critically evaluating most important issues of the surrounding society or culture of the students – a society and culture that are very much shaped by practices involving mathematics. In her conclusion, she emphasises that the ability to understand and to evaluate different practices of mathematics and the values behind has to be a component of Mathematical Literacy.

The demands and threats of *Knowledge Society* are referred to in most political declarations and justifications for educational policy. From an international or global point of view, this includes to investigate what approaches towards knowledge perceptions are taken in different countries, at the levels of policy and of practice; what are the most important knowledge conflicts at various social levels, and in particular in the educational systems, e.g. clashes between students' personal knowledge and the knowledge presented by teachers, between knowledge systems, between 'modern/popular' cultures and traditional cultures and between teachers' and students' views (see Clark, Keitel & Shimizu, 2006); and on the more general level, e.g. how are global technologies – especially the World Wide Web, television and print media – used to *promote or diminish diversity*, or what *effects of inequality* are reproduced.

The question how mathematics is perceived and used in *political debates and decision-making processes in particular about mathematics education* is a necessary complement to still be studied extensively (Gellert, Jablonka & Keitel, 2001): We have started case studies to investigate which connection is established between results of comparative studies on mathematical competencies and the attributions of causes and effects deduced from them in the public debate and to collect and analyse which criteria for political decisions and forms of decision-making processes are defined and stated, which kind of controlling mechanisms to secure quality is foreseen or used and on what the credibility of results is based, in particular in the media. We try to reconstruct the origin and history of such studies and confront criteria and decisions for selecting the participating institutions and experts, contrasting the official publications of national and international projects and the reconstruction of views and conceptions held by participating experts in interviews.

The history of the social reception of these studies is to be interpreted in the light of conflicts of interests and different interest groups that are identified by an analysis of published statements of all stakeholders in political decision processes, of representatives of interest groups in industry and economy, of teachers and of interpretations of public statements provided by mathematicians and experts in the ICT area. The interpretations of these statements in the light of the factual political interests are re-analysed on the base of the historical accounts of mathematics as means of social power and of the actual account of modern mathematics as a scientific discipline and technology provider, which finally leads to an answer, how much mathematics is needed to educate or create a well-informed and critical citizens for a democratic society.

References

Clarke, D.J., Keitel, C., & Shimizu, Y. (2006). *Mathematics classrooms in twelve countries: The insider's perspective*. Rotterdam: Sense Publishers.

Damerow, P., Lefévre, W. (Eds.). (1981). *Rechenstein, Experiment, Sprache. Historische Fallstudien zur Entstehung der exakten Wissenschaften* [Reckoning coins, experiment and language. Historical case studies on the beginning of the exact sciences]. Stuttgart: Klett-Cotta.

Damerow, P., Elwitz, U., Keitel, C., Zimmer, J. (1974). *Elementarmathematik: Lernen fuer die Praxis? Versuch der Bestimmung fachuebergreifender Curriculumziele* [Elementary mathematics: Learning for practice? An attempt to determine cross-curricular goals in mathematics]. Stuttgart: Klett.

Davis, P. (1989). Applied mathematics as social contract. In C. Keitel et al. (Eds.), *Mathematics, education and society* (Unesco Document Series No. 35, pp. 24–28). Paris: Unesco.

Davis, P. J., & Hersh, R. (1986). *Descartes' dream*. San Diego: Harcourt.

Gellert, U., Jablonka, E., & Keitel, C. (2001). Mathematical literacy and common sense in mathematics education. In B. Atweh et al. (Eds.), *Sociocultural aspects of mathematics education* (pp. 57–73). New York: Lawrence Erlbaum.

Howson, G. A., Keitel, C., & Kilpatrick, J. (1981). *Curriculum development in mathematics*. Cambridge: Cambridge University Press.

Jablonka, E. (2003). Mathematical literacy. In A. Bishop, K. Clements, C. Keitel, J. Kilpatrick, & F. Leung (Eds.), *Second international handbook of mathematics education* (pp. 77–104). Dordrecht: Kluwer.

Keitel, C., Kotzmann, E., & Skovsmose, O. (1993). Beyond the tunnel vision: Analysing the relationship between mathematics education, society and technology. In C. Keitel & K. Ruthven (Eds.), *Learning from computers: Mathematics education and technology* (pp. 242–279). Berlin: Springer.

Kline, M. (1985). *Mathematics and the search for knowledge*. Oxford: Oxford University Press.

Nissen, H., Damerow, P., & Englund, R. K. (1990). *Fruehe Schrift und Techniken der Wirtschaftsverwaltung im Vorderen Orient: Informationsspeicherung und -verarbeitung vor 5000 Jahren* [Early writing and technologies of the management and administration of economics in the Near East: Information storing and processing 5000 years ago]. Bad Salzdetfurth: Franzbecker.

OECD. (Eds.). (2000). *Programme for International Student Assessment (PISA)*. Paris: OECD.

Renn, J. (2002). *Wissenschaft als Lebensorientierung – eine Erfolgsgeschichte?* [Science as orientation for life – A story of success?]. Preprint 224. Berlin: Max-Planck-Institut für Wissenschaftsgeschichte.

Sierpinska, A., & Kilpatrick, J. (Eds.). (1998). *Mathematics education as a research domain: A search for identity*. Dordrecht: Kluwer.

Skovmose, O. (1994). *Towards a philosophy of critical mathematics education*. Dordrecht: Kluwer.

Snell, B. (1948). *Die Entdeckung des Geistes bei den Griechen*. [The discovery of reason by the Greeks]. Hamburg: Claasen & Goverts.

Woodrow, D. (2003). Mathematics, mathematics education and economic conditions. In A. Bishop, K. Clements, C. Keitel, J. Kilpatrick, & F. Leung (Eds.), *Second international handbook of mathematics education* (pp. 11–32). Dordrecht: Kluwer.

Chapter 9
Democratising Mathematics Education and the Role of Research

Alan J. Bishop

Abstract I have known Jeremy Kilpatrick since we were youthful participants at a UNESCO-sponsored seminar in Royaumont, France, on *New Trends in Mathematics Education* in 1971. Since then, we have been international colleagues, meeting at conferences or seminars, and later, we were editors at the same time of the two principal research journals in our field – the *Journal for Research in Mathematics Education* and *Educational Studies in Mathematics*. In all our working collaborations, I felt strongly that we shared a vision and a fundamental goal for mathematics education which I can best sum up with the word 'democratisation'. Always, Jeremy Kilpatrick was and is concerned to share his knowledge with others. In this paper, however, I analyse some of the current challenges to our ideas of democratisation through aspects of the curriculum, learners and learning, and teachers and teaching, together with promising areas of research.

Keywords Ethnomathematics • Numeracy • Mathematical literacy • Critical mathematics education • Classroom as social arena • Interdependence • Values (you choose)

Why Choose Democratisation as a Theme for This Chapter?

I have known Jeremy Kilpatrick since we were youthful participants at a UNESCO-sponsored seminar in Royaumont, France, on *New Trends in Mathematics Education* in 1971. We were invited to spend 2 weeks working in a 16-member group of international mathematics educators preparing Volume 3 of *New Trends in Mathematics Teaching* (Fehr and Glaymann 1972). Since then, we have been international colleagues, meeting occasionally at conferences or seminars, and less frequently for longer periods of collaborative study under sabbatical arrangements.

Later on, we were editors at the same time of the two principal research journals in our field – the *Journal for Research in Mathematics Education* and *Educational*

A.J. Bishop (✉)
Emeritus Professor, Faculty of Education, Monash University, Clayton, VIC 3122, Australia
e-mail: alan.bishop@monash.edu

© Springer International Publishing Switzerland 2015
E. Silver, C. Keitel-Kreidt (eds.), *Pursuing Excellence in Mathematics Education*, Mathematics Education Library, DOI 10.1007/978-3-319-11952-6_9

125

Studies in Mathematics, and we frequently discussed issues which we commonly faced as editors. The most common issue concerned the novelty and the relevance of the presented ideas. We were both acutely concerned that we were in essence responsible for the development of our field, for although we both had Editorial Boards to advise us, the 'buck' stopped at the editor's desk.

In all our working collaborations, I felt strongly that we shared a vision and a goal for mathematics education which I can best sum up with the word 'democratisation' – the making available to as wide a public as possible, and to as broad an education profession as possible, the ideas, the histories, the goals, and the many uses of mathematics.

Though he may not express it this way, I have no doubt that Kilpatrick portrays his striving for this goal through all his work – through his regular teaching and research at university (Athens, GA), through his publications, and through his conference lectures, seminars, and discussion groups. Over and above these 'normal' academic activities, he is a regular invitee to national and international committees and conferences, giving stimulating and challenging comment on the issues and topics of the moment. Always, he is concerned to share his knowledge with others, in an honest, free, and easy way – the ultimate democratiser.

Writing this paper gives me the opportunity not only to explore the ideas of democratising mathematics education, to analyse the obstacles to its progress, and as ways of overcoming those obstacles but also as a way of celebrating the work of Jeremy Kilpatrick.

Democratisation and Mathematics Education: Some Preliminary Thoughts

Democracy is a goal for many peoples in diverse societies at this time. We have seen thousands of people taking to the streets of their major cities, exercising their democratic right to free speech, and challenging their governments and leaders to address the many inequalities in their societies. Indeed, such is the fervour with which they long for what they believe democracy can bring them, in increasing numbers of cases they are dying in this quest. Democracy is such an iconic idea.

It may seem to be a big step from the incredible public and private traumas of creating more democratic societies to the more intellectual tasks of developing better mathematics education, but I see these tasks as running parallel. As part of the quest for more democratic societies, there is the demand for more democratic education, which means here a demand for much greater mathematical knowledge in all citizens. The essential challenge for mathematics educators is how to provide such quality mathematics education.

For these as well as other reasons, I am focusing this paper on the issue of democratisation in mathematics education, because it seems to be central to so many of our concerns in the field today. Also in this paper, I want to consider what research offers, because research trends are often good indicators of what our best thinkers believe are the crucial avenues to be explored in the future. Like Kilpatrick, I have always believed that research in mathematics education is fundamental to the development of our field and that it provides our principal intellectual protection against the narrow-minded and populist demands of many of our current educational and noneducational politicians.

At present, for example, in many countries, the economic rationalists continue to hold political power and their demands are for the efficient delivery of a narrow range of specified mathematical competencies by limited, and preferably inexpensive, means. I have no doubt that the goals being reached by this kind of approach in no way address the concerns for democratisation of mathematics education to which many of us aspire.

However, and fortunately, there are other approaches, other developments, and other possibilities for practice, and research trends can indicate what some of these are. In particular, I want to consider research that is challenging the conceptual obstacles that stand in the way of democratisation, by exploring the possibilities of other constructs and concepts.

I believe in emphasising here the issue of democratising mathematics education for three main reasons. Firstly, it is an aspect of education that is always present, but only rarely explicitly considered. The whole process of education is, for me, a democratising one, although in so many ways, the current versions seem to be more obstacles than solutions. Nevertheless, I see the main goal of mathematics education to be one of democratising mathematical knowledge by making it available, and understandable, to as many teachers, learners, policymakers, and others as possible.

Secondly, mathematics seems to be an easy subject in which to develop elites. We can see this in the way that mathematics has been used, and is still being used, to select students for higher education study. For example, in my country, if you study more and harder mathematics courses at school, this can give you more points towards your university entrance scores. These enable you to select a more prestigious subject there, e.g. medicine, and then you don't have to study anymore mathematics. We can see how necessary it is to overcome this elitist obstacle by the way that most adults believe that they were not able to succeed at mathematics and therefore have considerable admiration for anyone who does manage to succeed! It is a subject that has thrived on elitism throughout the ages and if mathematics educators are to achieve their democratising goals, then they will have to work even harder to overcome this elitist image.

Thirdly, the future of the world depends on the quality of the education we give to all our children. They *are* the future of the world in fact, and the fight against ignorance is one of the biggest challenges of our times. In our field, the task is the democratising of mathematical knowledge, without which the majority of our young people are being impoverished and disenfranchised. Sadly, mathematics is a subject that most people still believe they fail at or that they have little interest in. With the present approaches, it is easy for anyone to see when they have reached their own 'level of incompetence'.

A democratic education reflects on the knowledge from the past, from the perspective of the present, and with a vision of the future. Without the knowledge from the past, there is nothing with which to educate the present generation – there is merely the sharing of ignorance. Without a vision for the future, there is no goal towards which the education of the present learners should be directed – there is merely training.

The educational vision that I believe is needed is one that is democratising in the fullest sense – empowering, informative, action oriented, locally based but globally aware, reflective, critical, creative, and responsible. The questions though for this paper are what are the obstacles in the way of creating such a democratising mathematics education and how can research help to overcome the obstacles?

Mathematics Curricula

I think of our field of mathematics education research as being made up of a trio of basic constructs, each of which is surrounded by its social, political, and cultural context variables. In the next three sections, I will examine each of these in relation to our search for a democratic mathematics education The three constructs that form the essential foci of our research are:

- Mathematics curricula – involving aspects of content, sequences of ideas, and relationship to other topics, other subjects, and other contexts, both real and virtual
- Mathematics learning – involving characteristics of learners, types of learning, attitudes, beliefs, motivations, feelings, valuing, ways of remembering, imagining, and representing
- Mathematics teaching – involving interactions, explaining, clarifying, linking with other knowledge, inspiring, leading, and communicating

Mathematics curricula certainly rely on the knowledge from the past. Mathematics has a long history and mathematics curricula are locked into that history. Some knowledgeable critics argue that current curricula are too historical and can act as obstacles to democratic progress. Not only do they not reflect the ways that mathematics has changed over the centuries, but they also do not reflect the ways that society has changed.

To what extent do our mathematics curricula offer a democratic vision for the future? In my opinion, they offer the students very little of any vision for the future. Perhaps, if computers and Web-based activities are used a great deal, then we can see a high-tech vision being offered. But many of the ICT programmes on offer merely reproduce what was done in pre-technology days. There are some new and exciting virtual reality–based programmes available, but they are few and far between and are only rarely used in mainstream mathematics classrooms.

What is more of concern is that even if these exciting new programmes are available, they are often excluded from the typical mathematics programme promoted by our politicians and bureaucrats. As I indicated above, many of our educational politicians today have only a supposedly 'rosy' economic vision to offer the people and thus want mathematics education to guarantee a mathematical competence appropriate to the job the young person will do, that is, if they will have a job! They prefer training to education, and this view means that competence is valued over comprehension.

Looking at the issues of mathematics curricula from a research perspective, however, there are various democratising trends which are worth supporting and publicising to the wider public and professional audiences. For example, the idea of *ethnomathematics* has at its heart the recognition of the plurality of mathematical constructs and therefore of the need to move away from the hegemony of the Western model. The research challenges here are twofold: Firstly, how to meld the ethnomathematics ideas with the curriculum structures created by the need to teach hitherto only the Western mathematics model (Barton 1996). Secondly, there is considerable debate regarding ethnomathematics developments especially coming from sociological theorists as to whether that first challenge is an appropriate way to consider and democratise mathematics education (Gutstein 2003). For example, it brings with it a danger of presenting marginalised school learners with a localised curriculum, while their real need is to gain access to the mainstream culture of academic mathematics.

Another approach to democratisation through curricula is illustrated in the first international handbook on mathematics education, where Kilpatrick (1996), who edited Part 1 on Curriculum, argued that 'the mathematics curriculum, at least at both the school and university levels, has made two strong shifts in emphasis...' (p. 7) and 'away from an emphasis on abstract structures towards efforts to include more realistic applications, with an emphasis on the ways in which mathematics is used in daily and professional life' (p. 7). Also, he stressed 'there is a tendency for new curricula today to be organized around problems rather than topics' (p. 8). Hence, the *realistic mathematics* movement offers another democratising approach (de Lange 1996).

Another trend is that research is overcoming the conceptual obstacle of highly abstract content by considering the role of *context* instead. While studies certainly continue to examine issues of content, there is a growing interest in the context within which the mathematical practice is situated. Indeed, the construct of *mathematical practice* is also attracting a great deal of attention. Much of this work is based on the challenge coming from early studies of mathematical activities occurring outside school (e.g. Abreu 1995; Nunes et al. 1993) where the characteristics of the practice are markedly different from those inside the school. This phenomenon is well known in the field of adult and vocational education.

There is also an increasing research interest in *numeracy*, reflecting both a concern that mathematics teaching is not succeeding and also a desire to have a more relevant and context-related mathematics curriculum in the schools. Interestingly, as soon as educators begin to consider numeracy, in contrast to mathematics, there is an immediate local or community perspective brought into consideration, with little expectation that every community should have the same numeracy curriculum.

Developing numeracy, or *matheracy* as D'Ambrosio (1985) prefers to call it, is clearly a democratising goal that should be supported by all. This development relates also to the idea of *mathematical literacy*, which seems to push mathematics education towards a language image – but it is clear that mathematics does have its own literacy, and literacy issues, and this all helps to focus on what we could think of as 'hidden' obstacles for learners, particularly those from L2/3 backgrounds (Jablonka 2003). This construct is broader than just language, however, and as with the more general construct 'literacy', it refers chiefly to the political and social action required in a fully democratic society to further social justice goals through mathematics teaching (Gutstein 2003). This together with other constructs helps to further the idea of *critical mathematics education* (Skovsmose 1994; Brantlinger 2011).

The challenge from these studies is firstly how to determine the curriculum now that it is clear that simplistic ideas of 'knowledge transfer' are irrelevant and how to relate knowledge in one context to knowledge in another context. I would argue that local mathematical knowledge and practices should help to shape the local school mathematics curriculum, because there are often great differences between situations, knowledge, languages, and social practices in different parts of many countries. One must also not forget that differences of values are apparent between countries and cultures. Thus, I am always surprised that one can find so many similarities between mathematics curricula in different countries or even the identical curriculum. That is not the case with other subjects, with, for example, science beginning to localise its curricula more.

Mathematics Learners and Learning

One of the most accepted constructs from mathematics education in common use is also one of its most anti-democratic. It is the idea of *mathematical ability* and it is used by teachers, by learners, by parents, and by employers and was a construct that was popular in research many years ago, although it is now no longer of interest to most researchers. This is because it was associated with a description of learners *as if* it had no relation to any other descriptive variable.

It still is a kind of identifying label that learners almost carry around with them. Teachers prize their own ability at picking out the 'good' ones from the 'bad' ones. Once labelled in this way, learners find that they are then categorised for the purposes of grouping between classes and often for grouping within classes. Moreover, these different classes usually study different curricula, with 'proper' mathematics reserved for the 'good' ones. The 'not so goods' typically have a simplified curriculum emphasising simple routines and uninteresting tasks. If you are a 'good', 'able', or 'strong' student, you are directed to the harder subjects and probably will be encouraged to go to university to study yet more mathematics. If you are a 'poor' or 'weak' student, you are discouraged from doing anymore mathematics than is necessary, and you probably will drop it as soon as you can – the 'drop out' curriculum.

What is wrong with the construct of 'mathematical ability' and what are researchers now using and exploring? The problems are that it takes no heed of the social or curricular context in which the learner is learning, nor does it reflect the richness of individual learners (Gutierrez 2012). It assumes an overall high or low level of intellectual accomplishment no matter what the curriculum is, no matter who the teacher is or what methods he/she uses, and no matter what the age or the knowledge of the student is nor his/her motivation level, economic circumstance, state of health or mind, etc.

In contrast, research has increasingly shown the importance of the idea of *situated cognition* which describes the fact that when you learn anything, you learn it in a certain situation (Gutierrez 2012). You won't necessarily be able to transfer that learning to any other situation. This kind of research shows us what our common sense tells us, which in one situation you can appear clever but in another you can appear stupid. Change the classroom situation and the learner appears to be at a different ability level. In terms of mathematics education, we know that if the curriculum, teacher, or teaching method change, then other things can happen and the learner can often appear to be working at a much higher or a much lower (ability) level. For example, if the curriculum emphasises problems and issues of special interest to the learner, the level of performance can positively change. If the teacher uses methods that involve imaginative and creative activities, then other students begin to shine, not just the usual 'stars'. If the teacher switches from a routine manipulative approach in algebra to one that is more investigative, then other students begin to show promise. If the teacher and school adopt a social justice curriculum approach, with a more sociological flavour, then yet other students may thrive.

What researchers are far more interested in these days is the richness and diversity of students, which in research terms explores the idea of *mathematical abilities* (see Krutetskii 1976), a text which Kilpatrick edited from the original Russian. This is a construct which enables the description of learners to focus on the different ways that learners use to approach and solve new problems of learning. This kind of construct encourages teachers to look for the 'richness' in their students, not to narrowly categorise them as 'good' or 'weak' and not teach them in a way that ensures that they live up to, or down to, their (both teacher's and learner's) simplistic expectations. Looking for the richness in their students will support and encourage teachers to use other teaching approaches and to find curricular ideas that will develop and promote that rich diversity. Diversity in students is something to be welcomed and developed, not ignored or to be overcome. Democratic teaching recognises and celebrates difference and diversity.

Another general research approach that is having a democratising effect is that of thinking about *learners* rather than 'learning'. Time was when a focus on the psychology of mathematics learning was the main area of study, and researchers studied errors in learning, steps in learning, successful and unsuccessful learning, and different kinds of learning. While many useful ideas came from that research, the problem for the classroom is that teachers do not deal just with learning but predominantly with learners.

Learners are unique individuals existing in their specific social contexts. As research on situated learning developed and as the more social aspects of learning have come into prominence, the *social situation of learners* has been analysed to search for other explanations of successful or unsuccessful performance (Bishop and Forgasz 2007). Researches on the failure of bilingual learners in monolingual classrooms, or of farmers' children studying a totally urban-centred mathematics curriculum, or of physically or mentally handicapped learners, all help to shed light on other explanations of failure besides the attributes of the learners themselves.

The classroom is also a *social arena* where many roles are played out and where *significant others* can influence the quality of learning very directly. Learning mathematics should be a pleasurable as well as a challenging activity, carried out in stimulating contexts. If desired, learners should be able to collaborate with each other and learn from each other's viewpoints. Differences should be aired and compared constructively, not to see who is 'right' but to enable all the learners to learn from contrasting ideas and to learn why some ideas are more useful than others.

The new millennium is about *interdependence,* and young people will need to learn the skills of collaborating and of working constructively and sensibly with each other. The classroom of rows of desks where no one is allowed to talk, and everyone must study in silence for the whole lesson, or worse still just listen to the teacher, should be things of the past. Classrooms like these have no place in a democratic mathematics education. All they do is reinforce existing myths and obstacles to successful learning.

Teaching and Teachers of Mathematics

Already in the above two sections, we can see several ideas of relevance to the teaching of mathematics in more democratic ways. Ideas about different mathematical practices can make teachers more aware of relevant and valid mathematical activities that the students may participate in outside school, and the teacher should be trying to understand more about these.

Awareness of other mathematical practices enables teachers to create problems and tasks in class that can allow the students to demonstrate the knowledge that they have acquired elsewhere. This encourages the students to realise that they already know and practise mathematical ideas and skills, and this raises their self-esteem and encourages them even further. Teachers need to develop their skills in creating and working with learners in a collaborative and interdependent environment, rather than a competitive and teacher-centred one. *Culturally responsive pedagogy* is growing in interest as a consequence of research such as that on ethnomathematics.

In particular, in countries that are not based in the Western tradition, teachers could well make contact in their teaching with the other mathematical traditions of relevance to their students. Hopefully, as national curricula respond to the ever-increasing need to reflect their own nation's cultural ideas and values, teachers will have more opportunity to inject other and more relevant perspectives into their teaching. This can only benefit learners who at present struggle to master the 'alien' mathematics that is often a cultural product imposed by a colonial power (D'Ambrosio 1985). There are also many different teaching opportunities occurring within a numeracy or matheracy curriculum as long as the teachers have the training and expertise to be able to use them productively and as long as the mathematics education system creates the space for teachers to do this (Knijnik 1998).

The development of different kinds of constructs such as mathematical abilities in students has gone hand in hand with the idea of *flexible teaching*. This idea has derived mainly from work in distance education and with the realisation that merely having correspondence-type teaching at a distance is not adequate. Since the advent of the World Wide Web, it has become possible to incorporate all kinds of learning resources into the distance mode of teaching. The term 'flexible learning' describes this kind of system and although I don't like describing a teaching approach by using the 'learning' word, I sympathise with the idea that learners can by this means be more in control of their own learning. The idea has also come from those who teach adults, where there is a strong argument for learners having more control of their learning.

This idea is now coming to be realised as important within the school context also made possible by more information technology becoming available in schools. By this means, it makes the teacher think more about learning resources and opportunities for the learners, and thus, it extends the range of methods and approaches available to the teacher. In fact, perhaps the biggest development towards democratising the teaching of mathematics has been the move from considering 'teaching methods and processes' to ideas about 'learning resources and approaches'. This development changes the role of the teacher dramatically, and if it were to become widely accepted, it would surely have a strong democratising effect on mathematics education.

Perhaps, it seems that I am blaming teachers for all the errors and for all the anti-democratic teaching that goes on in mathematics classrooms. That is certainly not my intention, and I am only too aware of the roles others in the mathematics education system need to play if these democratic ideas are to be developed. Teachers are not autonomous and therefore, it is important that those of us who are in a position to challenge the anti-democratic obstacles surrounding mathematics teaching at present accept our share of the responsibility.

Democratising Mathematics Education and the Teaching of Values

Finally, at the heart of any discussion about democracy in mathematics education lies an area which I have been concentrating on for several years, namely, the issue of 'values'. This is a problematic area because firstly, we have little idea about what values mean in mathematics education. Secondly, there is very little consensus about what currently happens with values teaching/learning in mathematics classrooms or why. Thirdly, we know hardly anything about how potentially controllable such values teaching can be. Finally, it seems that many mathematics teachers do not even consider that they are teaching values when they teach mathematics. Changing that perception may prove to be one of the biggest hurdles to be overcome if we are to move to a more democratic mathematics education (Bishop et al. 2003).

We can see the order of the problem when we consider the seemingly simple question of whether mathematical values can actually be taught. What seems initially sensible is that values can be, and are, learnt by the students, just as they can presumably be learnt by anyone. But somehow to talk about *teaching* values suggests an intentionality, an instrumentality, about it – that it is just like one can teach subtraction or equation solving. For many educators, this does just not 'feel' right.

My perspective on this area comes from my consideration of mathematics as a form of cultural knowledge (Bishop 1988). As soon as one takes that perspective, then values come immediately into play, and in the book mentioned, I proposed six sets of values which I argue have come from the history of the development of mathematics. The subject has been developed by 'mathematicians' through the ages, and in my six-category analysis, I tried to unearth the values that have created and developed the special form of cultural knowledge that we know today as 'Western mathematics' – a hegemonic misnomer since many societies have contributed to its development. However, much more research needs to be done. I assume that values in mathematics education are the deep affective qualities which education fosters through the school subject of mathematics. However, only rarely does one find explicit values teaching in mathematics classrooms, one reason being the widespread belief that mathematics is a value-free subject. Skovsmose's (1996) chapter in the first *International Handbook of Mathematics Education* addressed values and democratic education explicitly, when he argued that:

> Critical mathematics education is concerned with the development of citizens who are able to take part in discussions and are able to make their own decisions. We therefore have to take into consideration the fact that students will also want, and should be given opportunities to 'evaluate' what happens in the classroom. This turns the focus on students' interests. (p. 1267)

This comment echoes the idea that for values education in mathematics to develop, there is a necessity to ensure that the mathematics classroom is a place of choices and of choosing. As well as making their own choices constantly, teachers should be presenting students with activities which encourage them to make choices, for example, about the selection of problems to be solved, about the solution approaches to be taken, about the criteria for judging the worth of solutions, and about the wider appropriateness of the mathematical models being taught.

It should be a natural part of the teacher's repertoire to present activities which require choices to be made: for example, a task such as 'Describe and compare three different proofs of the Pythagorean theorem' would inevitably engage students in discussing the values associated with proving. Even the simple act of presenting different problem-solving solutions to be compared and contrasted by the students stimulates the ideas of choice, criteria, and values. What Skovsmose's focus on students' interests does is to remind us that rather than thinking of mathematics teaching as just teaching mathematics to students, we are also teaching students through mathematics. They *are* learning values through how they are being taught, and in that sense, they are being educated rather than just being taught. This is why values are so important.

The acceptability of these ideas will of course depend ultimately on the capacity of teachers to engage with the idea of values and for teacher training professionals to take the idea of values seriously in their courses. For example, when choices are offered to the students and made by them, how do teachers respond? Do in fact teachers know what values they are currently implicitly teaching in the ways they respond to students' ideas? Perhaps, only when teachers give students more choices will they themselves be faced with responses which are new to them and which will therefore require them to become more aware of their own values.

This area of values is one which is fundamental not just to research but also to the whole notion of democratic mathematics education and it needs to be thoroughly investigated by both teachers and researchers. The results of such investigations would do much to enlarge our understandings of why mathematics teachers teach in the ways they do, of how to educate mathematically our future citizens, and of what are desirable, and feasible, goals for mathematics education in democratic societies as we move into the new millennium.

Finally, we need to constantly critique and evaluate our research approaches and recognise their limitations. Kilpatrick (1992) wrote in a landmark historical perspective: 'Educational researchers were not satisfying the requests of practitioners that they provide useful information, they were not satisfying funding agencies and their colleagues in the scientific disciplines that their work was valuable, and they were not satisfying their own expectations for what research should be. Far from living in a golden age, they seemed to be entering a depression' (p. 31). Is this still true today? More recently, Kilpatrick (2003) recorded, in an analysis of several standard-based research projects, that 'Curriculum developers and evaluators appear to be caught between politicians, on the one hand, who are mandating that research be done to evaluate programmes before they are implemented, and school people on the other hand, who are required to give tests for other purposes than research and who may consider participating in a research project a luxury they can't afford' (p. 487). There is no doubt that researchers in mathematics education work in a challenging profession. Nevertheless, I believe we have a responsibility to the future generations, who will inherit this complicated, and dangerous, world, to overcome the ignorance and fear of failure that is currently present in much mathematics education, by striving always to democratise our ideas, our theories, and our practices.

References

Barton, B. (1996). Making sense of ethnomathematics: Ethnomathematics is making sense. *Educational Studies in Mathematics, 31*, 201–233.

Bishop, A. J. (1988). *Mathematical enculturation: A cultural perspective on mathematics education*. Dordrecht: Kluwer/Springer.

Bishop, A. J., & Forgasz, H. J. (2007). Issues in access and equity in mathematics education. In F. K. Lester Jr. (Ed.), *Second handbook of research on mathematics teaching and learning.* Charlotte: Information Age Publishing.

Bishop, A. J., Seah, W. T., & Chin, C. (2003). Values in mathematics teaching. In A. J. Bishop, M. A. Clements, C. Keitel, J. Kilpatrick, & F. K. S. Leung (Eds.), *Second international handbook of mathematics education* (pp. 717–765). Dordrecht: Kluwer/Springer.

Brantlinger, A. (2011). Rethinking critical mathematics: A comparative analysis of critical, reform and traditional instructional texts. *Educational Studies in Mathematics, 78*(3), 395–411.

D'Ambrosio, U. (1985). Ethnomathematics and its place in the history and pedagogy of mathematics. *For the Learning of Mathematics, 5*(1), 44–48.

de Abreu, G. (1995). Understanding how children experience the relationship between home and school mathematics. *Mind, Culture, and Activity, 2,* 119–142.

De Lange, J. (1996). Using and applying mathematics in education. In A. J. Bishop, K. Clements, C. Keitel, J. Kilpatrick, & C. Laborde (Eds.), *International handbook of mathematics education* (pp. 49–97). Dordrecht: Kluwer/Springer.

Fehr, H., & Glaymann, M. (1972). *New trends in mathematics teaching* (Vol. 3). Paris: UNESCO.

Gutierrez, R. (2012). Context matters: How should we conceptualize equity in mathematics education? In B. Herbel-Eisenmann, J. Choppin, D. Wagner, & D. Pimm (Eds.), *Equity in discourse for mathematics education* (pp. 17–33). Dordrecht: Springer.

Gutstein, E. (2003). Teaching and learning mathematics for social justice in an urban, Latino school. *Journal of Research in Mathematics Education, 34*(1), 37–73.

Jablonka, E. (2003). Mathematical literacy. In A. Bishop, M. A. Clements, C. Keitel, J. Kilpatrick, & F. K. S. Leung (Eds.), *Second international handbook of mathematics education* (pp. 77–104). Dordrecht: Kluwer/Springer.

Kilpatrick, J. (1992). A history of research in mathematics education. In D. A. Grouws (Ed.), *Handbook of research on mathematics teaching and learning* (pp. 3–38). New York: Macmillan Publishing Company.

Kilpatrick, J. (1996). Introduction to section 1. In A. J. Bishop, K. Clements, C. Keitel, J. Kilpatrick, & C. Laborde (Eds.), *International handbook of mathematics education* (pp. 7–9). Dordrecht: Kluwer/Springer.

Kilpatrick, J. (2003). What works. In S. L. Senk & D. R. Thompson (Eds.), *Standards-based school mathematics curricula, what are they, what do students learn?* (pp. 471–489). Mahwah: Lawrence Erlbaum Associates, Publishers.

Knijnik, G. (1998). Ethnomathematics and political struggles. *Zentralblatt fur Didaktik Der Mathematik, 98*(6), 188–194.

Krutetskii, V. (1976). *The psychology of mathematical abilities in schoolchildren* (J. Kilpatrick, Ed. & I. Wirszup, Trans.). Chicago: University of Chicago Press.

Nunes, T., Schliemann, A., & Carraher, D. (1993). *Street mathematics and school mathematics.* New York: Cambridge University Press.

Skovsmose, O. (1994). *Towards a philosophy of critical mathematics education.* Dordrecht: Kluwer/Springer.

Skovsmose, O. (1996). Critical mathematics education. In A.J. Bishop, K. Clements, C. Keitel, J. Kilpatrick, & C. Laborde (Eds), *International handbook of mathematics education* (pp. 1257–1288). Dordrecht: Kluwer.

Chapter 10
On the Diversity and Multiplicity of Theories in Mathematics Education

Pearla Nesher

Abstract The paper starts with a brief history of theories that guided mathematics education research presented in PME meetings and publications. There seems to be a "feeling" that mathematics education should rely on theoretical grounds to become a legitimate field of research. Yet, the current situation with multiplicity of theories is troubling. Whenever a new perspective emerges, it becomes a fashion, and many researchers followed and praised this new paradigm, abandoning the "old" paradigms. On the basis of Niss's attempt to define the concept of theory, the paper examines the "constructivist" vs. the "realist" theories and suggests that under the school of constructivism as practiced in mathematical education research, multiplicity of theories is unavoidable. A claim is made that it is time to examine the option of having theories in mathematics education that may possibly be verified or falsified.

Keywords Constructivism • Realism • Theories • Math education • Peirce

Preamble

I dedicate this paper to my dear colleague Jeremy Kilpatrick whom I have known
   since 1976 and of whom my appreciation keeps growing. In 1976, the second
   IGMPE meeting took place in Germany at "Haus Ohrbeck," a charming place
   near Osnabruck. The group was small, about 60 people.
The most impressive intellectual part of this meeting was Jeremy Kilpatrick's plenary
   lecture about (if my memory does not betray me) problem-solving. He gave his next
   PME plenary lecture in Montreal (Kilpatrick 1987). This lecture about "construc-
   tivism" became a cornerstone in the agenda of mathematics education research.
I was the president when PME decided to publish a book in the ICME Study Series
   about accumulated research on mathematics and cognition. As a member of the
   ICME international committee, Jeremy joined me in that enterprise which

P. Nesher (✉)
Department of Mathematical Education, The University of Haifa, Haifa, Israel
e-mail: nesherp@edu.haifa.ac.il

© Springer International Publishing Switzerland 2015
E. Silver, C. Keitel-Kreidt (eds.), *Pursuing Excellence in Mathematics Education*, Mathematics Education Library, DOI 10.1007/978-3-319-11952-6_10

137

climaxed with Jeremy and I sitting in an Arabic restaurant in Acre, Israel, finalizing some details about the book: Jeremy's touch made that book (Nesher and Kilpatrick 1990).

My great appreciation rests on Jeremy Kilpatrick's amazing ability to articulate his thoughts and opinions about the most burning issues in the field of mathematics education. He has served mathematics education for about 50 years: studying in the years of writing the "new math" curriculum (SMSG), teaching mathematics, bringing to attention the Soviet writings about mathematics education, serving as JRME editor, summarizing and updating research on problem-solving, writing on the history of math education research, discussing the "math wars," "adding it up." etc. He is up to date with research as well as with politics; he broadly encompasses issues of theory and practice. I could go on, but I want to stress that Jeremy always expressed his deep reflections with elegance and modesty, avoiding antagonism even with those with whom he argues. For me, he serves as a model.

Introduction

In the late 1970s, constructivism movement was spreading among math education community through undoubting faith and not too much questioning. At the 1987 PME in Montreal, Jeremy Kilpatrick raised some questions and doubts concerning constructivism in an effort to initiate a rational discussion. Typically, he was raising theoretical questions that could not be easily dismissed. He argued that many constructivist pedagogical ideas and practices were already rooted in other theoretical and pedagogical practices (Kilpatrick 1987).

One of his central criticisms concerned epistemology and ontology. These remain bothersome issues for mathematical education community. For example, Sriraman and English's theoretical volume expresses irresolution and direction seeking, rather than an answer to the current disturbing situation (Sriraman and English 2010). There seems to be a "feeling" that mathematics education should rely on theoretical grounds to become a legitimate field of research. Yet, the current situation is troubling as Arzarello writes: "This situation [*diversity of theories*, P.N] is unpleasant and may create a sense of unease since it marks a strong difference with respect to the paradigm of the disciplines like mathematics" (Arzarello 2010, p. 507).

The Search for a Theory of Mathematics Education

The need for research in mathematical education becomes obvious in the failure of so many students to pass mathematics. Despite good intentions, teachers' efforts, and volumes of research studies, we really have no theory or knowledge of how to change those upsetting results.

The quest for theory in mathematics education is not new (Kilpatrick 1981; Steffe and Nesher 1996). I will sketch its brief history. It is usually said that the search for theories started with the establishment of JRME or with the foundation of PME. However, mathematics education as a field grew out of the "Sputnik era," accompanied by theories of curriculum (Bruner 1960) and attempts to do research on curriculum development (School Mathematics Study Group (1958–1977), Minnemath (1961–1970), The Nuffield Project (1960–1970)).

Research in the 1970s and the 1980s leaned mainly on cognitive and developmental psychology as was manifested in the establishing of IGPME (International Group for the Psychology of Mathematics) in 1976. Exposure to Piaget's theory (Piaget 1950, 1952, 1970; Piaget et al. 1960; Piaget and Inhelder 1969) and the translation of Soviet research in mathematics education (Kilpatrick and Wirszup 1969) led to the rise of the "constructivist" movement in the USA, which was crystallized in the theoretical terms of "radical constructivism" (von Glaserfeld 1984, 1991) and accompanied by "teaching interviews" in the pedagogical arena (Cobb and Steffe 1983; Steffe and Thompson 2000).

Radical constructivism kept the basic principle of constructivism that the learners have to construct and elaborate their own knowledge, which does not come "ready-made" from the outside world, emphasizing cognitive experience and *individual* development. Radical constructivists, however, went further, presenting an epistemological stand that omitted ontology altogether and replaced *truth of a theory* with *viability* (see below).

Soon, this theoretical stance seemed insufficient and was replaced, again with enthusiasm, by "social constructivism" which rests on Vygotsky's theory (Vygotsky 1962/1934, 1978/1935). Ernest, who articulated this philosophy, asserted that conventions of human language, rules, and agreements are the basis for establishing and justifying the truths of mathematics (Ernest 1991, 2006, 2010).

However, the quest for theories for research in mathematics education did not stop there. Whenever a researcher has realized that it enlightened part of the process of teaching and learning, a new perspective became a fashion and many researchers followed and praised this new paradigm, abandoning the "old" paradigms. The volumes of PME proceedings evident these shifts in focus, from "radical constructivism" to "social constructivism," "ethno-mathematics," "situated learning," "the theory of didactic situations" (Brousseau 1984, 1997), "class interaction" (Yackel 1996; Steinbring 2008), "interactionism" (Bauersfeld 1988), "discourse analysis" (Jungwirth 1993), "semiotics" (Gellert 2010), "signs and gestures" (Working session 3, PME31, 2007), "semiotic-cultural, anthropological activity theory" (Eisenhart 1991; Cobb et al. 1996; Prediger 2004; Jungwirth 2010), "equity" and political issues (Keitel 2006), and more. The shifts were sweeping that even the name of the group became a subject of hot debate.

Clearly, taking one theoretical approach can mean searching one aspect and ignoring many others, such being the nature of research in any field. Different theories can describe the same phenomena at different levels. For example, physical reality is examined at one level (or aspect) by "high-energy physics," on another by "solid-state physics," or another example by "geography" or "oceanography" and so forth. Likewise, the complex reality of mathematics learning and teaching probably requires a variety of perspectives. Different aspects of mathematics education should be studied at different levels with different paradigms. The problem lies not in the *diversity* of theories, but rather in clarifying the *different levels of the phenomena* under study; defining each level clearly then allows us to avoid *multiplicity* of theories, by converging on appropriate common assumptions, terms, and methods. By *diversity* of theories, I mean different theories aimed at different phenomena; by *multiplicity,* I mean, several theories all dealing with the same phenomena, but none falsified by any other.

Putting aside the meta-theoretical arguments which constitute the core of Sriraman and English's important volume, I raise the question of what will make a theory a fruitful resource for our research.

What We Expect from a Theory?

I begin with Niss's attempt to define the concept of theory (Niss 2007):

A *theory* is a system of concepts and claims with certain properties, namely:

- A theory consists of an *organized system of concepts* (including ideas, notions, distinctions, terms etc.) and *claims* about some extensive domain, or class of domains, consisting of objects, processes, situations, and phenomena.
- In a theory the *concepts are linked in a connected hierarchy* (oftentimes, but not necessarily, of a logical or proto-logical nature), in which a certain set of concepts, taken to be basic, are used as building blocks in the formation of the other concepts.
- In a theory the *claims are either* basic hypotheses, assumptions, or axioms taken as *fundamental* (i.e., not subject to discussion within the boundaries of the theory itself) or statements obtained from the fundamental claims by means of *formal* (including deductive) *or material* (i.e., experiential or experimental with regard to the domain) *derivations*. (Ibid, P. 1308)

[All *italics* by Niss]

Niss mentions six different purposes that theories serve (for full exposition, see Niss 2007, pp. 1308–1309):

(a) To provide *explanation* of some observed phenomenon [this one is mentioned as crucial].
(b) To provide *prediction* of the (possible) occurrence of certain phenomena as a claim resulting from the (possible) fulfillment of the preconditions of their occurrence.
(c) To provide *guidance for action or behavior.*
(d) To provide *a structured set of lenses*, through which aspects or parts of the world can be approached [mentioned as most important].
(e) To provide *a safeguard against unscientific approaches* to a problem, issue or theme.
(f) Protection against attacks from outside.

[All *italics* by Niss]

These purposes are problematical: entry (f), for example, is not intrinsic to the theory but rather relates to outside agencies. As to entry (a), we know that, in principle, there may be different explanations for the same phenomena. How do we choose among them? How do we know which explanation is true? An explanation depends on the theory and is derivative from it. Entry (c) is also a derivative of belief in a certain theory. Entry (e) raises the controversial issue of what is considered a scientific or unscientific theory, an issue at the heart of arguments, attitudes, and judgments about theories of mathematics education (both inside and outside our community).

There remain entries (b) and (d). Entry (b) is considered to be an intrinsic purpose of a scientific theory: *to predict possible occurrences*, I would add to it the requirement that a theory should also predict what *will falsify it*. Entry (d) seems fruitful for understanding and explaining many phenomena and is considered as most important goal of current research in mathematics education. However, the consequences of (b) and (d) are different and perhaps incompatible (see below).

Peirce's Pragmaticism

Niss's description of a theory comprises two distinct domains: *the claims of a theory* (the beliefs) on one hand and *the observed phenomena* on the other (see above). Such a view brings in the question of *truth* – the relation between a proposition (claim) and a phenomenon. We base our beliefs and actions on claims that we accept as true. This position is reminiscent of the realist school in philosophy, as exemplified by Russell in his simplified version (Russell 1912):

> It will be seen that minds do not *create* truth or falsehood. They create beliefs, but when once the beliefs are created, the mind cannot make them true or false,… …What makes a belief true is a *fact*, and this fact does not (except in exceptional cases) in any way involve the mind of the person who has the belief. (Ibid, p. 82)

This theory assumes arriving at *truth by correspondence:* the status of a *claim* (proposition) being in agreement with a *fact* (reality). Traditionally, some in math education community (Vinner and Tall 1982; Gold 1999; Dorfler 2002) have seen realism and Platonism as indistinguishable; yet, in modern philosophy, several versions of realism accept neither Plato's nor Russell's version. All believe in the existence of a real world (reality) but differ on how the cognized human can bridge between a man-made theory and that real world.

One such realist theory is Charles S. Peirce's "Pragmaticism" (not to be identified as "pragmatism," elaborated by James and Dewey, or "pragmatic realism," by Putnam (1987, 1990)).[1] Peirce has been cited by many researchers in math education, mainly because of Peirce's semiotic system (Dorfler 2002; Otte 2006; Radford 2006; Steinbring 2006). I would like to stress his kind of realism, which Weiner,

---

[1] Peirce himself labeled his own logical formulation *pragamaticism* in order to distinguish it from James's psychological version of pragmatism (Weiner 1958).

who edited selected papers by Peirce (Weiner 1958), named "Commonsense Realism":

> The gist of that message lies in Peirce's profound sense of fallibility and yet supreme value of honest, persevering inquiry by individual minds sharing a common desire to learn and a common faith that an indefinite community of such investigators must sooner or later discover the truth and the reality corresponding to it. (Ibid, p. VIII)

Peirce was aware that our beliefs are achieved by a community of investigators yet are independent of the world, of external things with qualities that we experience. Thus, beliefs, we think to be true, might clash with our experience. At the same time, he also realized the impossibility of arriving at absolute truth (that he called the Truth (with capital T), a never-ending quest at which we strive.

In particular, two of Pierce's papers – *The Fixation of Belief* (Peirce 1878/1958) and *How to Make Our Ideas Clear* (Peirce 1877/1958) – are relevant to our concerns, especially, insofar as they deal with inquiry. He elaborated on beliefs and doubts and described how one starts an inquiry: "The irritation of doubts causes a struggle to attain a state of belief. I shall call this struggle *inquiry*" (ibid, p. 99). "Beliefs guide our desires and shape our actions" (ibid, p. 98). In the positive sense, they direct our habits. After criticizing beliefs that are fixed by tenacity, authority, and apriority, Peirce suggests that "the question of validity is purely one of fact and not of thinking" (ibid. p. 95), and later he writes:

> To satisfy our doubts, therefore, it is necessary that a method should be found by which our beliefs may be caused by nothing human, but by some external permanency – by something upon which our thinking has no effect. (Ibid. p. 107)

Peirce claims that *fallibilism* and *reference* are essential to a proper conception of truth, but he is also aware that beliefs are inherent in our cognition. He suggests a sign system that makes it possible to find and define the relationship between our cognition and reality.

The realist epistemology in its various versions goes back to the Galileo's doctrine that the scientist aims at a true explanation of observable facts. It includes different forms of "realism" like Popper's epistemology who claimed that we never arrive at true theory; we are lucky to obtain refutation of our conjectures: "Our falsifications thus indicate the points where we have touched reality as it were" (Popper 1989/1963, p. 116).

There are, however, completely different theories of knowledge (epistemologies) that do not require the correspondence to an independent world. Constructivism is one such approach.

Social Constructivism

Constructivism, like realism, encompasses a diversity of theories (cf. Ernest 1991, 2010 for full discussion). All have in common (a) that knowledge is not passively received, it is actively constructed by the subject on the basis of experience and earlier constructions (knowledge), (b) knowledge (theory) serves to structure experience and not to search the ontology of a real world, and (c) the aim, in attaining knowledge, is achieving *viability* and not *truth*. Here, they depart from Piaget's Genetic Epistemology.

Constructivism thus views all of our knowledge as "constructed"; it does not reflect any external realities (as correspondence theories might hold). Rather, perceptions of truth are viewed as contingent on convention, human perception, and experience. The best known variants of this epistemology are *radical constructivism* and *social constructivism*. Because Kilpatrick's paper (Kilpatrick 1987) dealt extensively with radical constructivism, I will elaborate more on social constructivism

Social constructivism holds that truth is constructed by social processes, it is historically and culturally specific, and that it is in part shaped through the power struggles within a community.

As Ernest reflects:

> …based on the seminal work of Vygotsky, Social constructivism regards individual learners and the realm of the social as indissolubly interconnected. Human beings are formed through their interaction with each other as well as by their individual processes…mind is viewed as social and conversational. (Ernest 2010, pp. 43–44)

In Ernest (ibid.) writings, language and signs play major roles in social constructivism. Thus, theories in mathematics education which fall under the titles "class discourse," "discourse analysis," and "interactionists" are variants of social constructivism.

From this point of view Sfard's (2008), theory of "commognition" seems also to fall within social constructivism. In Sfard's analysis, objects are creations of discourse: "This [object construction]. would usually involve many interrelated acts of reification, saming, and encapsulation …" (ibid, p. 182). Thus,

> Reification consists in associating of a noun with a discursive process and replacing narratives about those processes with equivalent narratives about the object signified by the noun. (Ibid, p. 192)

In addition, the process of encapsulation replaces talk in the plural with talk of the singular.

The purpose of the theories that fall under the umbrella of social constructivism is expressed in Niss's (2007) purpose (d): "To provide *a structured set of lenses*, through which aspects or parts of the world can be approached" (ibid, p. 2008). The epistemologies "realism" and "constructivism" are behind Niss's purposes (b) and (d), respectively, but these now appear to have incompatible implications.

Theories in Mathematics Education, Reconsidered

It is important and practical to distinguish the epistemological background of various theories. Different epistemological require different controls in planning research and summarizing its findings. Research in the tradition of social constructivism mostly reflects on phenomena, interpreting them according to its specific framework, be this "interactionist," "semiotic," or "commognition."

Most research in math education falls within the scope of social constructivism. In general, a situation is given (usually a protocol of a conversation) and the researcher uses his or her theoretical glasses to examine the situation, typically, succeeding in showing that the particular theory helps to understand and explain the situation (the phenomenon). Nothing can contradict or falsify the explanation, because the phenomenon and its explanation share a common (not independent) ground. Or, more exactly, any researcher can explain the same situation from a different perspective (Gellert 2010). Therefore, all theories in the framework of social constructivism continue to live side by side, resulting in the present state of affairs. In fact, constructive research exactly accords with purpose (d) in Niss's paper (Niss 2007): they provide *a structured set of lenses*, through which aspects or parts of the world can be approached and explained.

Networking of theories suggested by several authors (Bikner-Ahsbahs and Prediger 2010; Gellert 2010; Jungwirth 2010) does not change the fact that the combined theories aim to interpret given phenomena without raising the main question, namely, how can we make sure that these coordinated theories are true and that the others exploring the same phenomena are false? If deciding among theories is impossible, and if we have no clear criteria to select among theories, beyond fashion, authority, or "being nice" to each other, we are doomed to remain with a *multiplicity* of theories.

The realist school of thought, in all its different variants, has its own criteria. Claims of a given theory are judged to be true or false according to the facts (phenomena) they observe.[2] One could argue that this is true only for the natural sciences. But there are linguistic theories that rise or fall when examining, for example, a corpus of languages.

It is time that we examine this possibility of verification or falsification of theories in mathematics education. There are clear phenomena which we can observe in the processes of learning and teaching mathematics. We observe these phenomena at different levels: the individual learners and the cognitive schemes that enable them to acquire specific contents of mathematics on different levels. the discourse in the classroom, the teacher as an expert who moderates and participates in that discourse, the political aspects and forces that intervene in the process, and more. Each one of these aspects should be studied at its level, hence the justification for *diversity* of theories.

The struggle to curtail the *multiplicity* of theories for the *same* phenomena deserves an honest effort. "Easier to say than to do," one might think. However, a possible starting point, for theorizing on any level, lies at being surprised, startled by something, while we are in touch with an observed phenomena (be they the outcomes of a certain strategy or competing strategies, students' failure in a task thought to be easy, an unexpected discourse on a given problem, etc.). Such a

---

[2] An actual example: Physicists at CERN look for a particle named "Higgs boson." If they do not find it, the standard theory of physics falls.

reference (ontology) on the basis of which one theorizes can serve as an independent measure of theoretical *claims*. This means a lot: trying to agree on genuine doubt that creates disquiet (worry), converging on decided and fixed terms, suggesting testable claims formulated in a manner that might become true or false, and so on.

I admit that such a suggestion does not accord with most current views and studies in mathematical education research.[3] Yet, as Pring says:

> Once one loses one's grip on 'reality, or questions the very idea of 'objectivity', or denies a knowledge-base for policy and practice, or treats facts as mere inventions or construction, then the very concept of research seems unintelligible. (Cited by Dahl 2010)

Pring's statement stands as a rebuttal to Radford's (2006) epistemological musing that:

> Here, we abandon the idea of Truth in the essentialist metaphysical tradition, according to which Truth is that which remains once all that is ephemeral has been removed – an idea that goes back to Plato's aristocratic ontology (see Radford, 2004). We also abandon the idea of objectivity as an uncompromised access to transcendental entities. (p. 60)

In principle, there are two alternatives for research in mathematics education. First, we could continue with the current tradition of various constructivist theories that reject any form of realism and are content with "deepening" and "understanding." This option is expressed by Lester (2010):

> I suggest that rather than adhering to one particular perspective, we act as *bricoleurs* by adapting ideas from a range of theoretical sources to suit our goals – goals that should aim not only to deepen our fundamental understanding of mathematics learning and teaching, but also to aid us in providing practical wisdom about problems practitioners care about. (p. 83)

Cobb also suggests working in a bricolage manner, employing whatever is available to theorize in mathematics education research (Cobb 2007). Lerman (2010) adds: "I am not surprised by the multiplicity of theories in our field and the debates about their relative merits, nor do I see. It as a hindrance" (p. 108).

Taking the first road may enrich our intellectual points of view but will not solve the issue of multiple theories. It includes no way to stop annexing more and more theories (even in a network) that interpret and "understand" phenomena in mathematics education and no way to decide among them.

Our second option is to embrace the ambition of formulating testable and fallible theories that can be judged according to their truth value: not Truth as described by Radford in a dogmatic hyperbole, but "truth" in the spirit of Peirce's Pragmaticism. Peirce emphasizes fallibilism and reference as essential to a proper conception of truth as approximation, incompleteness, and partiality.

---

[3] It is interesting to read survey results published in 2009 by PhilPapers Surveys, most believed theories: According to a survey of professional philosophers and others on their philosophical views which was carried out in November 2009 (taken by 3226 respondents, including 1,803 philosophy faculty members and/or PhDs and 829 philosophy graduate students), 44.9 % of respondents accept or lean toward correspondence theories, 20.7 % accept or lean toward deflationary theories, and 13.8 % epistemic theories. http://philpapers.org/surveys/results.pl?affil=All+ respondents&areas0=0&areas_max=1&grain=medium

We need theories with which we could formulate claims (out of theoretical rationales) that predict some consequences so that they can be verified or falsified. These need not be grand theories, but they should state as a minimum under which conditions they will be true or false.

If we adhere to the second option, some current theories may vanish and others may survive, at least, until further inquiry will send them too to the trash basket of history.

References

Arzarello, F. (2010). Commentary on networking theories – An approach for exploring the diversity of theoretical approaches. In B. Srirman & L. English (Eds.), *Theories in mathematics education*. New York: Springer.

Bauersfeld, H. (1988). Interaction, construction, and knowledge – Alternatives for mathematics education. In T. C. D. Grouws (Ed.), *Effective mathematics teaching* (pp. 22–46). Reston: National Council of Teachers of Mathematics.

Bikner-Ahsbahs, A., & Prediger, S. (2010). Networking of theories - an approach for exploiting the diversity of theoretical approaches. In B. Srirman & L. English (Eds.), *Theories in mathematics education* (pp. 483–503). New York: Springer.

Brousseau, G. (1984). *The crucial role of the didactical contract in the analysis and construction of situations in teaching and learning mathematics*. Occasional paper. Bielefeld: FRG, Universitat Bielefeld, Institute fur Didaktik der Mathematik. 54.

Brousseau, G. (1997). *Theory of didactical situations in mathematics*. Dordrecht: Kluwer.

Bruner, J. C. (1960). *The process of education*. Cambridge: Harvard University Press.

Cobb, P. (2007). Putting philosophy to work: Coping with multiple theoretical perspectives. In K. F. Lester (Ed.), *Second handbook of research on mathematics teaching and learning* (Vol. 1, pp. 3–39). Charlotte: NCTM, Information Age Publishing.

Cobb, P., & Steffe, P. (1983). The constructivist researcher as teacher and model builder. *Journal for Research in Mathematics Education, 14*, 83–94.

Cobb, P., Jaworski, B., et al. (1996). Emergent and sociocultural views in mathematical activity. In L. P. Steffe & P. Nesher (Eds.), *Theories of mathematical learning* (pp. 3–21). Mahwah: Lawrence Erlbaum Associates.

Dahl, B. (2010). Commentary on the fundamental cycle of concept construction underlying various theoretical frameworks. In B. Sriraman & L. English (Eds.), *Theories of mathematical education* (pp. 193–206). New York: Springer.

Dorfler, W. (2002). Formation of mathematical objects as decision making. *Mathematical Thinking and Learning, 4*(4), 337–350.

Eisenhart, M. (1991). Conceptual frameworks for research. Ideas from a cultural anthropologist: Implications for mathematics education researchers. In *Proceedings of the thirteenth annual meeting of Psychology of Mathematics Education – North America*. Blacksburg: R. Underhill.

Ernest, P. (1991). *The philosophy of mathematics education*. London: Palmer.

Ernest, P. (2006). A semiotic perspective of mathematical activity: The case of number. *Educational Studies in Mathematics, 61*(1–2), 67–101.

Ernest, P. (2010). Reflections on theories of learning. In B. Sriraman & L. English (Eds.), *Theories of mathematical education: Seeking new frontiers* (pp. 39–47). New York: Springer.

Gellert, U. (2010). Modalities of local integration of theories in mathematics education. In B. Sriraman & L. English (Eds.), *Theories of mathematical education: Seeking new frontiers* (pp. 537–550). New York: Springer.

Gold, B. (1999). Social Constructivism as a philosophy of mathematic, by Paul Ernest; What is mathematics Really? by Reuben Hersh. *The American Mathematical Monthly, 106*(4), 373–380.

Jungwirth, H. (1993). Routines in classroom discourse. An ethnomethodological approach. *European Journal of Psychology of Education, 8*(4), 375–387.

Jungwirth, H. (2010). On networking strategies and theories' compatibility: Learning from an effective combination of theories in a research project. In B. Sriraman & L. English (Eds.), *Theories of mathematical education: Seeking new frontiers* (pp. 519–535). New York: Springer.

Keitel, C. (2006). Mathematics, knowledge and political power. In J. Maass & W. Schloglmann (Eds.), *New mathematics education research and practice* (pp. 11–23). Rotterdam: Sense Publishers.

Kilpatrick, J. (1981). The reasonable ineffectiveness of research in mathematics education. *For the Learning of Mathematics, 2*(2), 22–29.

Kilpatrick, J. (1987). What constructivism might be in mathematics education. In J. Bergeron, N. Herscovics, & C. Kieran (Eds.), *Proceeding of the international conference on the psychology of mathematics education* (Vol. 1, pp. 3–28). University of Montreal, Montreal.

Kilpatrick, J., & Wirszup, I. (Eds.). (1969). *Soviet studies in the psychology of learning and teaching mathematics*. Stanford: School Mathematics Study Group.

Lerman, S. (2010). Theories of mathematics education: Is plurality a problem? In B. Sriraman & L. English (Eds.), *Theories of mathematical education: Seeking new frontiers* (pp. 99–109). New York: Springer.

Lester, K. F. (2010). On the theoretical, conceptual, and philosophical foundations for research in mathematics education. In B. Sriraman & L. English (Eds.), *Theories of mathematical education: Seeking new frontiers* (pp. 67–83). New York: Springer.

Nesher, P., & Kilpatrick, J. (Eds.). (1990). *Mathematics and cognition. ICMI studies*. Cambridge: Cambridge University Press.

Niss, M. (2007). Reflections on the state and trends in research on mathematics teaching and learning from here to Utopia. In K. F. Lester (Ed.), *Second handbook of research on mathematics teaching and learning* (National Council of Teachers of Mathematics, Vol. 2, p. 1323). Charlotte: Information Age Pub.

Otte, M. (2006). Mathematical epistemology from a Peircean semiotic point of view. *Educational Studies in Mathematics, 61*(1–2), 11–38.

Peirce, C. S. (1877/1958). How to make our ideas clear. In P. P. Weiner (Ed.), *Charles S. Peirce: Selected papers (Values in universe of chance)* (pp. 113–137). New York: Dover Publications, Inc.

Peirce, C. S. (1878/1958). The fixation of our belief. In P. P. Weiner (Ed.), *Charles S. Peirce: Selected papers (Values in universe of chance)* (pp. 91–113). New York: Dover Publications, Inc.

Piaget, J. (1950). *The psychology of intelligence*. London: Routledge & Kegan Paul.

Piaget, J. (1952). *The child's conception of number*. New York: W. W. Norton & Company.

Piaget, J. (1970). *Genetic epistemology*. New York, NY, US Columbia University Press

Piaget, J., & Inhelder, B. (1969). *The psychology of the child*. New York: Basic Books.

Piaget, J., Inhelder, B., et al. (1960). *The child's conception of geometry*. New York: W.W. Norton & Company.

Popper, K. R. (1989/1963). *Conjectures and refutations: The growth of scientific knowledge*. London/New York: Routledge.

Prediger, S. (2004). Intercultural perspective on mathematics learning – Developing a theoretical framework. *International Journal of Science and Mathematics Education, 2*(3), 377–406.

Putnam, H. (1987). *The many faces of realism*. La Salle: Open Court.

Putnam, H. (1990). *Realism with a human face*. Cambridge, MA: Harvard University Press.

Radford, L. (2006). The anthropology of meaning. *Educational Studies in Mathematics, 61*(1–2), 39–65.

Russell, B. (1912). *The problems of philosophy*. Oxford University Press, UK (1967).

Sfard, A. (2008). *Thinking as communicating*. Cambridge: Cambridge University Press.

Sriraman, B., & English, L. (Eds.). (2010). *Theories of mathematical education: Seeking new frontiers. Advances in mathematics education*. New York: Springer.

Steffe, L. P., & Nesher, P. (Eds.). (1996). *Theories of mathematical learning*. Mahwah: Lawrence Erlbaum Associates.

Steffe, L. P., & Thompson, P. (2000). Teaching experiment methodology: Understanding principles and essential elements. In K. R. A. Lesh (Ed.), *Handbook of research design in mathematics and science education* (pp. 267–306). Mahwah: Lawrence Erlbaum Associates.

Steinbring, H. (2006). What makes a sign? An epistemological perspective on mathematical interaction. *Educational Studies in Mathematics, 61*(1–2), 133–162.

Steinbring, H. (2008). *The construction of new mathematical knowledge in classroom interaction – An epistemological perspective*. New York: Springer.

Vinner, S., & Tall, D. (1982). Existence statements and constructions in mathematics and some consequences to mathematics learning. *The American Mathematical Monthly, 89*(10), 752–756.

von Glaserfeld, E. (1984). An introduction to radical constructivism. In P. Walzwlawick (Ed.), *The invented reality* (pp. 17–40). New York: Norton.

von Glaserfeld, E. (Ed.). (1991). *Radical constructivism in mathematics education*. Dordrecht: Kluwer.

Vygotsky, L. S. (1962/1934). *Language and thought*. Cambridge, MA: MIT Press and Wiley.

Vygotsky, L. S. (1978/1935). *Mind in society*. Cambridge, MA: Harvard University Press.

Weiner, P. P. (Ed.). (1958). *Charles S. Peirce: Selected writings (Values in a universe of chance)*. New York: Dover.

Yackel, E. (1996). *Social interaction and individual cognition*. Mahwah: Lawrence Erlbaum Associates.

Chapter 11
Toward a Profession of Mathematics Education: Guidance from Jeremy Kilpatrick's Words and Deeds

Edward A. Silver

Abstract With reference to a list of attributes of professions suggested by Shulman—the obligation of *service* to others, the need for *understanding* of a scholarly or theoretical kind, a domain of skilled performance or *practice*, the exercise of *judgment* under conditions of unavoidable uncertainty, the need for *learning from experience* as theory and practice interact, and a professional *community* to monitor quality and aggregate knowledge—this chapter considers how those attributes apply to the professional field we call mathematics education. Using specific examples of contributions Jeremy Kilpatrick has made in his writings and professional activities, the paper argues that his leadership by example has mapped a pathway along which mathematics education can advance toward achieving the status of a full-fledged profession.

Keywords Mathematics education • Profession/professional • International community • Research in mathematics education

I found the task of writing a chapter for this volume intended to celebrate Jeremy Kilpatrick's lifelong accomplishments to be both very easy and very difficult. It was easy because I respect and admire Jeremy and all that he has contributed to the field of mathematics education. I am also grateful for all that he has given to and done for me personally. Yet, the task was also quite daunting because there is so much that one could say, so many facets of Jeremy's contributions and accomplishments, that it was a difficult task to choose just one or two things to be the focus of a chapter. Moreover, as a former student of Jeremy Kilpatrick—someone who received extensive, careful, detailed, critical, formative feedback from him on many papers in the

Preparation of this paper was supported in part by the Usable Scholarship in Education (USE) Initiative, a project funded by the University of Michigan. Any opinions, conclusions, or recommendations expressed here are those of the author and do not imply institutional endorsement.

E.A. Silver (✉)
School of Education, University of Michigan, Ann Arbor, MI, USA
e-mail: easilver@umich.edu

© Springer International Publishing Switzerland 2015
E. Silver, C. Keitel-Kreidt (eds.), *Pursuing Excellence in Mathematics Education*, Mathematics Education Library, DOI 10.1007/978-3-319-11952-6_11

past—I felt a strong desire to create a chapter that was both beautifully written and intellectually substantive. I view these as distinctive characteristics of Jeremy's writing, and I know that he steadfastly sought to instill them also in me through his cajoling and his critique. Though I recognize that this chapter falls short of the Master's mark, I do wish to add my voice to that of the other contributors to this volume in addressing some noteworthy aspects of Jeremy's professional work.

It was my good fortune to have had Jeremy Kilpatrick as a professor and mentor when I began my studies in mathematics education more than 40 years ago. I arrived at Teachers College, Columbia University, in the fall of 1972 with the intention of enrolling in a course of study that would both lead to a Master's degree in mathematics education and satisfy the requirements set by New York State for certification as a teacher of mathematics in grades 7–12. I satisfied that goal and then continued on to earn my doctoral degree in mathematics education. When I arrived at Teachers College, I knew almost nothing about mathematics education or even about education more generally (beyond what I had experienced myself).

As an undergraduate, I had studied mathematics but not education. After graduating from college, I accepted a fellowship to study for a PhD in mathematics at the University of Maryland. During my first year in the PhD program, I met Professor James Fey, who held a joint appointment in the Mathematics Department and in the College of Education at the University of Maryland and who invited me to attend a mathematics education seminar that he was teaching.

When family circumstances forced me to curtail my doctoral studies in mathematics and return to New York, Jim Fey suggested that I investigate the mathematics education program at Teachers College, where he had received his doctorate a few years earlier. I recalled that advice later after returning home and being hired to teach at a school in New York City. A close friend of mine taught at this school, and there was no teacher on staff with a strong background in mathematics, so the school was willing to hire me even though I was neither trained to be a teacher nor certified to do so. But a condition of my employment was that I take courses and work toward meeting teacher certification requirements. That fortuitous set of circumstances allowed me to meet Jeremy Kilpatrick and to begin what has been a journey of more than four decades.

When one thinks of Jeremy Kilpatrick, many words and phrases are likely to be evoked, such as scholar, editor, mentor, critic, teacher, expert, colleague, mathematics educator, world traveler, gentleman, and professional. I have chosen the final word in that list—*professional*—to be the focus of this chapter.

Professionals and Professions

What does it mean to be a professional? The word has many meanings in English. According to the Merriam-Webster dictionary, professional refers to a person "engaged in one of the learned professions" or a person "characterized by or conforming to the technical or ethical standards of a profession" (http://www.merriam-webster.com/dictionary/professional). As we see, these characterizations of a professional depend on the meaning of the word profession, which also has many meanings. That same dictionary defines a profession as "a calling requiring specialized knowledge and often long and intensive academic preparation" (http://www.merriam-webster.com/dictionary/profession). Thus, to call someone a professional, in this sense of the word, is to make a claim both about the person and the practice in which that person engages on a regular basis.

To call Jeremy Kilpatrick a professional then seems to imply that mathematics education is a profession. Yet, Jeremy has written on several occasions that mathematics education is not a profession. For example, in one of his most recent papers, when discussing the relationship between mathematics and mathematics education, he asserted: "We can think of mathematics and mathematics education as partners and as complements. Mathematics is both a profession and a discipline, whereas mathematics education is neither. It is a field of practice and field of study" (Kilpatrick 2013, p. 10). In this succinct phrasing, he echoes a refrain from an earlier paper of his on mathematics education as an academic field: "In my view, mathematics education has not attained the status of a discipline, and it is not completely a profession" (Kilpatrick 2008, p. 36).

Although Jeremy may have concluded that mathematics education is not a profession, it is my contention that his record of work and accomplishment as a mathematics educator embody key features of professionalism that can serve as a paradigm case of what it would mean for mathematics education to be not only a field of study and practice but also a profession.

As Lieberman (1956) noted, "there is no authoritative set of criteria by means of which we can distinguish professions from other occupations" (p. 1). Yet, it is possible to identify some generally accepted characteristics. There are many such lists, and Shulman (1998, p. 516) offers a short list of attributes of professions that I find useful:

- The obligation of *service* to others, as in a "calling"
- *Understanding* of a scholarly or theoretical kind
- A domain of skilled performance or *practice*
- The exercise of *judgment* under conditions of unavoidable uncertainty
- The need for *learning from experience* as theory and practice interact
- A professional *community* to monitor quality and aggregate knowledge

Jeremy Kilpatrick: The Professional Mathematics Educator

Jeremy Kilpatrick's career to date illustrates several of the items in Shulman's list of professional characteristics. Consider, for example, the obligation of service to others. In a career that he began in 1952 as a junior high school teacher of mathematics and science at Garfield Junior High School in Berkeley, California, and that he continues to this day as Regents Professor of Mathematics Education at the

University of Georgia, Jeremy Kilpatrick has given generously of his time and expertise in service to others.

Jeremy has compiled an extensive record of work with major educational organizations in the United States, such as the National Council of Teachers of Mathematics, the College Board, the American Educational Research Association, and the Mathematical Association of America. He has worked on major national and international assessments, such as the National Assessment of Educational Progress, Third International Mathematics and Science Study (TIMSS), International Mathematical Olympiad, and the Programme for International Student Assessment (PISA). And he has been a key contributor to the International Commission on Mathematical Instruction (ICMI) internationally and in the United States to many endeavors of the National Science Foundation, the National Research Council, and the National Academy of Education. His record epitomizes what it means to respond to an obligation of service to others.

Jeremy's mathematics education contributions related to Shulman's categories of scholarly understanding of a theoretical kind and a domain of skilled performance are numerous and known well to readers of this volume, so I will not try to enumerate them all here. The list of topics to which he has made important scholarly contributions would surely include mathematical problem-solving, mathematics ability, mathematical proficiency, school mathematics curriculum, and mathematics assessment. Moreover, he is widely and justifiably regarded as a seminal figure in the domain of research in mathematics education because of his editorship of the *Journal for Research in Mathematics Education* for 6 years; his contributions to numerous influential handbooks, reports, and committees; and his writings on research that have offered valuable critique and guidance to the field.

In Jeremy's writing, one finds a passionate commitment to the value of research in mathematics education and an equally passionate criticism of research that is not done well. For example, in his 1981 paper entitled *The Reasonable Ineffectiveness of Research in Mathematics Education*, he decries the lack of attention to theory in mathematics education research, and he argues that greater attention to theory would both improve the scientific quality of research in the field and increase the relevance of the research to educational practice. That paper (Kilpatrick 1981) is typical of many that Jeremy has written, offering on the one hand a stinging critique of the current state of research in mathematics education and on the other sound guidance about what might be done not only to improve the intellectual and technical quality of research but also to increase its relevance of teachers, curriculum developers, and teacher educators.

Though one cannot make causal attribution on the basis of available evidence, there can be little doubt that the situation has changed in mathematics education research since Jeremy's 1981 paper. Several prominent commentaries on research in mathematics education have noted a shift toward greater attention to theory in the field (e.g., Lerman 2010; Lerman et al. 2002; Sfard 2005; Silver and Herbst 2007). For example, Silver and Herbst (2007) note:

> Although many legitimate questions and concerns can and should be raised about the 'state of the art' in using theory in our field, a perusal of articles in major research journals in our field reveals that theory is alive and well in recent work, as evidenced by the frequent

appearance of the words theory, theoretical framework, and theorizing. Further support for
this claim comes from an interesting analysis by Lerman and Tsatsaroni (2004) of articles
published in two major research journals in our field (*Educational Studies in Mathematics*
[ESM] and the *Journal for Research in Mathematics Education* [JRME]). They noted a
sharp decline between 1990–1995 and 1996–2001 in the number of published articles that
they judged to have no evident theoretical framing; the decline was from 10% to 5% for
ESM and from 24% to 11% for JRME. (p. 41)

Interestingly, some 25 years after Kilpatrick's lament about the lack of attention
to theory, one sees a shift to a new concern about the proliferation of different theories used by researchers in mathematics education (e.g., Bikner-Ahsbahs and
Prediger 2006)!

Returning to Shulman's list of characteristics of a profession, we find another that
is robustly evident in Jeremy Kilpatrick's career, namely, the need for a professional
community to monitor quality and aggregate knowledge. Several features of
Jeremy's activities are noteworthy in this regard. He has long served in a role that is
analogous to that of a "shuttle diplomat," representing and explaining the American
perspective on mathematics teaching and learning to colleagues in other countries
and bringing to the attention of the American audience the perspectives and work of
non-US mathematics educators. His commitment to an international community of
mathematics educators is evident in his writings (e.g., Kilpatrick 1993, 2008); his
extensive work with ICMI, TIMSS, and PISA; as well as his editing of numerous
volumes of *Soviet Studies in Mathematics Education*, his visiting faculty appointments abroad (including in Colombia, England, Germany, Italy, Spain, and Sweden),
and his engagement with numerous international projects, boards, and conferences.

Jeremy has also worked persistently and diligently to ensure that the community
of mathematics education does not exclude mathematicians. As noted earlier, one
finds in many of his writings attention to the relationship between mathematics and
mathematics education (see, e.g., Kilpatrick 1981, 2008). Jeremy is, of course, not
alone among mathematics educators in addressing this important issue, but he has
written eloquently on the subject, and he has also been attentive to this connection
in his professional life. He has been an active participant in the work of the
Mathematical Association of America—a professional association of mathematicians—
serving two terms on its Board of Governors and as a member of several committees
and projects. Particularly noteworthy in this regard is his leadership of a National
Research Council study committee on mathematics learning that produced a highly
influential report, *Adding It Up: Helping Children Learn Mathematics* (Kilpatrick
et al. 2001), that was praised for its careful treatment of core issues and relevant
evidence by combatants on both sides of the so-called math wars in the United
States.

One more aspect of Jeremy's commitment to community merits special mention,
namely, his insistence on the inclusion of mathematics teachers as partners with
researchers and scholars in mathematics education. In my opinion, throughout his
illustrious career, Jeremy has remained faithful to his roots as a mathematics teacher.
He taught in junior high school for only a few years, but his appreciation for the
work of classroom teaching never faltered. In graduate school, I witnessed Jeremy
Kilpatrick, the mathematics teacher, when I took his courses in non-Euclidean

geometry and in mathematical problem-solving. I recall that he was a careful, skillful lecturer who encouraged active participation on the part of students, invited and carefully answered their questions, posed thought-provoking questions to stimulate student learning, and provided meticulous feedback on homework assignments. He was, of course, equally adept in teaching courses in which the content was less mathematical and more educational, but I had the impression that he truly enjoyed the occasions in those classes when the conversation turned to the interplay between specific mathematical topics and the broader issues of teaching and learning.

In several of Jeremy's papers, he addresses directly the important role that mathematics teachers can play as members of the mathematics education community. For example, in his 1981 paper on the ineffectiveness of mathematics education research, he wrote: "If researchers in mathematics education are to become effective in improving the practice of mathematics teaching, they should develop a stronger sense of community, which would include practicing teachers as collaborators in research" (Kilpatrick 1981, p. 28).

Beyond Jeremy's words expressing the need for a mathematics education community that embraces both scholars and teachers, two personal experiences stand out as illustrations of how this commitment permeated his work in tangible ways. The first came during my years as a graduate student; the second came some years later.

As noted earlier, I came to my graduate studies as a novice not only in mathematics education as an academic field but also in mathematics teaching as a skilled practice. I recall puzzling over complex problems of practice and seeking wisdom and guidance from the books and papers I read in my courses, occasionally being inspired by ideas I found there but often feeling a sense of despair at the absence of practical assistance and actionable proposals. As my academic advisor, Jeremy was subjected to many hours of my ruminations during his office hours at Teachers College and in letters that we exchanged (in the days before e-mail!) while he was at Cambridge, England, as a visiting professor. I never recall Jeremy becoming impatient with my fumbling attempts to develop as a teacher and as a scholar. Instead, I recall that he was incredibly supportive of my questioning and my quest, always available to provide moral support, a sympathetic ear, and a suggestion of something else that I might want to think about or read. And I have good reason to think that I was not unique in receiving this patient support from Jeremy; several of my fellow graduate students have shared similar stories about their interactions with Jeremy. Though he was well known as a demanding critic of our written papers, he was equally known as someone willing to listen in a supportive manner to our practical questions and concerns.

The second example came several years after completing my doctorate when I had the good fortune to collaborate with Jeremy on some projects that afforded further glimpses of his commitment to forming a community of mathematics educators that included teachers in a central way.

In 1989, I was approached by the College Board to prepare a book on mathematics teaching that would be part of a series intended to display innovative ideas about teaching secondary school subjects (history, world languages, science, mathematics) in ways that were both intellectually stimulating and accessible to a wide range of students. At the time, Jeremy was completing his term as chair of the Mathematical Sciences Academic Advisory Committees of the College Board, an organization in the United States uniquely positioned to address academic issues that lie at the interface between secondary school and university education.

I agreed to take on the task with the condition that I could invite coauthors to work with me. My coauthors were Jeremy and Beth Schlesinger, a talented secondary school teacher from San Diego, California. I knew Beth from my time on the faculty at San Diego State University. She was well known in the region as an exceptional mathematics teacher, someone adept at stimulating the most gifted mathematics and equally successful in reaching students with far less apparent ability or interest.

I was uncertain about how the collaboration would work because Beth and Jeremy did not know each other, and they had what appeared to be quite divergent spheres of activity. But I need not have been concerned because Jeremy and Beth quickly developed a deeply respectful appreciation for the contribution that each could make to the book. Jeremy embraced Beth's insights into mathematics teaching and helped to connect them to his own ideas and experiences, as well as to relevant scholarship. The book produced through this collaboration (Silver et al. 1990) offers one example of what it might be possible for a community of practitioners and scholars to produce.

A few years later, I served with Jeremy and a number of other educators on a committee of the National Board for Professional Teaching Standards that was charged with the task of developing a set of standards for judging highly accomplished practice in teaching mathematics at the upper elementary/lower secondary school level. The composition of the committee was quite diverse, including a number of teachers with quite diverse teaching assignments and mathematics backgrounds. Thus, I was in a position to witness firsthand Jeremy's respectful, collegial manner when interacting with other committee members. He listened carefully and added his own ideas by building on the contributions of others in the group; when he disagreed, he did so forcefully, yet respectfully. He quite naturally emerged as a leader within the group not only because he had wonderful insights to contribute but also, and perhaps more importantly, because he was able to synthesize the contributions of other group members to create a clear, coherent, and cogent product.

Toward a Profession of Mathematics Education: Lessons from Jeremy Kilpatrick

In my view, Jeremy Kilpatrick's insistence on the inclusion of teachers as members of the mathematics education community goes well beyond simply advocating that we should "play well with others." In his exhortation, I see a core connection to another of Shulman's characteristics of a profession: the need for learning from experience as theory and practice interact. Moreover, I think it signals the centrality of teaching as a core concern of mathematics education.

It might be possible to simplify the pathway to mathematics education as a profession if one restricted the community of mathematics educators to include only individuals with university appointments. But doing so would also inflict a great cost by weakening the profession's ties to teaching as a core concern. In a paper prepared in celebration of the 100th anniversary of the International Commission on Mathematics Instruction, Jeremy argued for the centrality of teaching (Kilpatrick 2008). He argued by analogy from Jens Høyrup's claim regarding the discipline of mathematics: "Teaching is not only the vehicle by which mathematical knowledge and skill is transmitted from one generation to the next; it belongs to the essential characteristics of mathematics to be constituted through teaching" (Høyrup 1994, p. 3). Jeremy applied this idea to mathematics education: "In a similar fashion, one can argue that an essential characteristic of mathematics education is to be constituted through teaching—in this case, through teaching teachers as well as students, and through teaching teachers to teach as well as to understand mathematics" (p. 15).

The disconnection between educational practice, on the one hand, and the ideas and work of scholars in relevant academic fields on the other has drawn the attention of numerous commentators (e.g., Kennedy 1997; Lagemann 2000; Stein and Coburn 2010). Some have argued that disconnection is inevitable and not deeply problematic (e.g., Kerlinger 1977); others have contended that it is a major problem that weakens education—an unfortunate by-product of a complex set of circumstances, such as poor research quality, lack of training of practitioners in how to use research, inadequate funding, or weak infrastructure support for the translation and dissemination of research to practice (e.g., Burkhardt and Schoenfeld 2003). One thread running through many of the commentaries is the challenge of making research and scholarship not only respectable and high quality according to the criteria applied by academicians but also relevant and useful to practitioners.

Fenstermacher and Richardson (1994) argued that a "dichotomy between allegiance to the discipline and allegiance to the activity of education troubles all foundational studies of education, whether philosophy of education, history of education or the psychology of education" (p. 49). They discuss what they view as status-related pressures "to be disciplinary" within educational psychology, and they challenge educational psychologists to "deploy their disciplinary tools and techniques in a morally grounded search for better ways to educate rather than continuing to perfect tools and techniques within disciplinary boundaries and then sally forth to argue how education should conform to these improved concepts, theories, and research findings" (p. 53). If one substitutes the phrase *mathematics education* for educational psychology in their commentary, the issue is likewise posed for our field.

As Jeremy and I noted in a paper we prepared about two decades ago based on interviews with a number of prominent researchers in mathematics education:

> Diverse disciplinary allegiances have been formed to provide the theory and methods that undergird research in mathematics education. Some of the disciplinary connections, such as those to psychology and to mathematics itself, are longstanding; others, such as those to

anthropology and sociology, are of a more recent vintage. These disciplinary connections raise many issues, such as the feasibility and wisdom of fostering strong ties to other intellectual disciplines in a field like mathematics education that also needs strong ties to educational practice. We noted above the firm belief among those colleagues we interviewed that quality research in mathematics education needs to be solidly grounded simultaneously in the field of educational practice and in a theoretical framework tied to a scholarly discipline. This dual requirement characterizes a fundamental challenge for research in mathematics education ... One of the costs associated with such a disciplinary focus to research is that it can lose its connection to educational practice, or at least appear to have lost such a connection. Repeated attempts to build strong connections to foundational disciplines like mathematics and psychology may be one reason why mathematics education research has in the past been regarded with skepticism by educational practitioners, including not only precollege teachers but also colleagues who deal primarily with matters of curriculum development or teacher education in mathematics. (Silver and Kilpatrick 1994, pp. 739–740)

If mathematics education is to complete its journey toward becoming a profession, then Shulman tells us that we need to consider the interaction of theory and practice, and we need to learn from our examination of that interaction. What Jeremy Kilpatrick has reminded us often in his papers is that our chances of learning from the interaction of theory and practice are enhanced if the community of mathematics educators includes both teachers and scholars. Moreover, as Jeremy has also demonstrated through his actions throughout his career, it is important also to include mathematicians and to build a community that is truly international.

As noted earlier, Jeremy has written on several occasions that mathematics education is not yet fully a profession. Yet, in the brief summary of some aspects of his career that I have provided in this chapter, I think we can see that Jeremy Kilpatrick has marked a path for us to follow toward making mathematics education not only a field of study and practice but also a profession. His leadership by example suggests an amendment to an assertion he made in a recent paper he wrote on the topic of leadership in mathematics education:

Mathematics education, therefore, has leaders of at least two kinds: We have intellectual leaders who lead because of ideas, and we have what we might call political-social leaders who occupy leadership positions. There are many of the latter, and whether they come to take on genuine leadership roles is often a function of various and peculiar circumstances. (Kilpatrick 2013, p. 7)

Jeremy Kilpatrick's highly accomplished career exemplifies a third type of leader—blending the two categories he identifies and adding another dimension—leadership by example. These individuals lead through ideas and actions, words, and deeds, laying a path for others to follow. In his career, Jeremy has exemplified the characteristics of professionalism that mark a clear path for the rest of us toward a profession of mathematics education. We can all benefit from his exemplary leadership now and in the future, as we have in the past.

In his career, Jeremy has often played the important role of historian for mathematics education. In several of his papers, he has chronicled the development of the field of mathematics education, the formation of the international community of mathematics educators, the growth of mathematics education within the United States, and the development of research activity in general and in specific areas of inquiry (e.g., problem-solving). In so doing, he has helped us see and understand our evolutionary roots, to recognize critical moments and important influences in our history, and to appreciate the interconnectedness of the ideas and the human agents engaged in the activity. He modestly omits himself from the story line, but his role as a leader and shaper of mathematics education is indisputable. When the history of the development of mathematics education as a profession is written in the future, there is little doubt that Jeremy Kilpatrick will have a central role in the story.

References

Bikner-Ahsbahs, A., & Prediger, S. (2006). Diversity of theories in mathematics education—How can we deal with it? *Zentralblatt für Didaktik der Mathematik, 38*, 52–57.

Burkhardt, H., & Schoenfeld, A. H. (2003). Improving educational research: Toward a more useful, more influential, and better-funded enterprise. *Educational Researcher, 32*(9), 3–14.

Fenstermacher, G. D., & Richardson, V. (1994). Promoting confusion in educational psychology: How is it done? *Educational Psychologist, 29*, 49–55.

Høyrup, J. (1994). *In measure, number, and weight: Studies in mathematics and culture.* Albany: State University of New York Press.

Kennedy, M. (1997). The connection between research and practice. *Educational Researcher, 26*(4), 4–12.

Kerlinger, F. N. (1977). The influence of research on education practice. *Educational Researcher, 6*(8), 5–12.

Kilpatrick, J. (1981). The reasonable ineffectiveness of research in mathematics education. *For the Learning of Mathematics, 2*(2), 22–29.

Kilpatrick, J. (1993). Beyond face value: Assessing research in mathematics education. In G. Nissen & M. Blomhøj (Eds.), *Criteria for scientific quality and relevance in the didactics of mathematics* (pp. 15–34). Roskilde: Roskilde University, IMFUFA.

Kilpatrick, J. (2008). The development of mathematics education as an academic field. In M. Menghini, F. Furinghetti, L. Giacardi, & F. Arzarello (Eds.), *The first century of the International Commission on Mathematical Instruction (1908–2008): Reflecting and shaping the world of mathematics education* (pp. 25–39). Rome: Istituto della Enciclopedia Italiana.

Kilpatrick, J. (2013). Leading people: Leadership in mathematics education. *Journal of Mathematics Education at Teachers College, 4*(Spring–Summer), 7–14.

Kilpatrick, J., Swafford, J. O., & Findell, B. (Eds.). (2001). *Adding it up: Helping children learn mathematics.* Washington, DC: National Academy Press.

Lagemann, E. C. (2000). *An elusive science: The troubling history of education research.* Chicago: The University of Chicago Press.

Lerman, S. (2010). Theories of mathematics education: Is plurality a problem? In B. Sriraman & L. English (Eds.), *Theories of mathematics education* (pp. 99–110). New York: Springer.

Lerman, S., & Tsatsaroni, A. (2004, July). *Surveying the field of mathematics education research.* Paper prepared for Discussion Group 10 at the tenth international congress on Mathematical Education, Copenhagen. Accessed July 6, 2013, at http://myweb.lsbu.ac.uk/~lermans/ESRC ProjectHOMEPAGE.html

Lerman, S., Xu, G., & Tsatsaroni, A. (2002). Developing theories of mathematics education research: The ESM story. *Educational Studies in Mathematics, 51*, 23–40.

Lieberman, M. B. (1956). *Education as a profession.* Englewood Cliffs: Prentice-Hall.

Sfard, A. (2005). What could be more practical than good research? *Educational Studies in Mathematics, 58*, 393–413.

Shulman, L. S. (1998). Theory, practice, and the education of professionals. *Elementary School Journal, 98*, 511–526.

Silver, E., & Herbst, P. (2007). Theory in mathematics education scholarship. In F. K. Lester (Ed.), *Second handbook of research on mathematics teaching and learning* (pp. 39–67). Greenwich: Information Age.

Silver, E. A., & Kilpatrick, J. (1994). *E pluribus unum*: Challenges of diversity in the future of mathematics education research. *Journal for Research in Mathematics Education, 25*, 734–754.

Silver, E. A., Kilpatrick, J., & Schlesinger, B. (1990). *Thinking through mathematics*. New York: The College Board.

Stein, M. K., & Coburn, C. E. (2010). Reframing the problem of research and practice. In C. E. Coburn & M. K. Stein (Eds.), *Research and practice in education: Building alliances, bridging the divide* (pp. 1–13). Lanham: Rowman & Littlefield.

Jeremy Kilpatrick

Bio Sketch (in His Own Words)

Before joining the faculty at the University of Georgia in 1975, I taught at Teachers College, Columbia University. I hold A.B. and M.A. degrees from the University of California at Berkeley, M.S. and Ph. D. degrees from Stanford University, and an honorary doctorate from the University of Gothenburg. I was appointed Regents Professor of Mathematics Education at Georgia in 1993.

My publication activities include coediting the series Soviet Studies in the Psychology of Learning and Teaching Mathematics from 1969 to 1975, editing the *Journal for Research in Mathematics Education (JRME)* from 1982 to 1988, and editing the *JRME* Research Commentary section from 2004 to 2008. I edited the chapters on curriculum for the 1996 *International Handbook of Mathematics Education*, the chapters on research for the 2003 *Second International Handbook of Mathematics Education*, and the chapters on international perspectives for the 2013 *Third International Handbook of Mathematics Education*. I also coedited the 1998 publication *Mathematics Education as a Research Domain*, the 2003 publication *A Research Companion to Principles and Standards for School Mathematics*, and the 2003 publication *A History of School Mathematics*. My articles include a chapter on the history of research in mathematics education in the 1992 *Handbook of Research on Mathematics Teaching and Learning*, a coauthored research report on an innovative precalculus course in the 1996 Volume 3 of *Bold Ventures: Case Studies of U.S. Innovations in Mathematics Education,* and a 2012 article on the new math as an international phenomenon in *ZDM: The International Journal on Mathematics Education*.

I chaired the Committee on Mathematics Learning of the U.S. National Research Council, whose report *Adding It Up: Helping Children Learn Mathematics* was published in 2001. I also served on the RAND Mathematics Study Panel, whose report *Mathematical Proficiency for All Students: Toward a Strategic Research and*

© Springer International Publishing Switzerland 2015

E. Silver, C. Keitel-Kreidt (eds.), *Pursuing Excellence in Mathematics Education*, Mathematics Education Library, DOI 10.1007/978-3-319-11952-6

Development Program in Mathematics Education appeared in 2002. Both reports addressed the development of proficiency in teaching mathematics by improving teachers' knowledge, skill, and practice. Strengthening the professional education of mathematics teachers was the aim of the Center for Proficiency in Teaching Mathematics (CPTM, 2002–2007), in which I served as a principal investigator. CPTM was a collaborative Center for Teaching and Learning funded by the National Science Foundation that brought together the University of Georgia and the University of Michigan as research partners.

I have taught courses in mathematics education at several European and Latin American universities and have received Fulbright awards for work in New Zealand, Spain, Colombia, and Sweden. I was a charter member of the U.S. Mathematical Sciences Education Board and served two terms as Vice President of the International Commission on Mathematical Instruction. I am a Fellow of the American Educational Research Association, a National Associate of the U.S. National Academy of Sciences, and a member of the U.S. National Academy of Education. In 2003, I received a Lifetime Achievement Award from the National Council of Teachers of Mathematics and, in 2007, the Felix Klein Medal honoring lifetime achievement in mathematics education from the International Commission on Mathematical Instruction. My research interests include teachers' proficiency in teaching mathematics, mathematics curriculum change and its history, assessment, and the history of research in mathematics education.

In interviews by Dave Roberts for the *International Journal for the History of Mathematics Education* (2009) and by Henrique Guimarães for the *International Journal for Research in Mathematics Education* (2011), I discussed the profound influence of George Pólya on my life and work. Not only was he my teacher and a member of my dissertation committee at Stanford but I had the unique opportunity to serve as his assistant in courses for teachers and freshman seminars. One of the publications of which I am most proud is *The Stanford Mathematics Problem Book: With Hints and Solutions,* which I coauthored with him. Pólya taught me more than mathematics; he taught me what it means to be a gentleman and a scholar. He personified excellence.

Jeremy Kilpatrick: Summary of Professional Background and Accomplishments

Education and Professional Experience

Education

AA (mathematics and science), Chaffey College, 1954
AB (mathematics), University of California, Berkeley, 1956
MA (education), University of California, Berkeley, 1960
MS (mathematics), Stanford University, 1962
PhD (mathematics education), Stanford University, 1967

Professional Experience

Garfield Junior High School, Berkeley, CA, Mathematics and Science Teacher, 1957–1960
Teachers College, Columbia University, Assistant Professor of Mathematics, 1967–1970; Associate Professor of Mathematics, 1970–1975
University of Georgia, Professor of Mathematics Education, 1975–1993; Regents Professor of Mathematics Education, 1993–present

Visiting Appointments

University of Cambridge, England, Visiting Lecturer in Education, 1973–1974
Institut für Didaktik der Mathematik, Universität Bielefeld, Germany, Guest Professor,1976
Scuola Matematica Interuniversitaria, Perugia, Italy, Faculty Member, 1988

© Springer International Publishing Switzerland 2015
E. Silver, C. Keitel-Kreidt (eds.), *Pursuing Excellence in Mathematics Education*, Mathematics Education Library, DOI 10.1007/978-3-319-11952-6

Universitat Autònoma de Barcelona, Spain, Senior Lecturer in Education, 1989, 1991
Equity 2000 Mathematics Summer Institute, Nashville, TN, Faculty Member, 1991
Universidad de los Andes, Bogotá, Colombia, Guest Lecturer in Education, 1993
Göteborgs Universitet, Gothenburg, Sweden, Guest Professor in Didactics of Mathematics, 1993

Selected Professional Awards/Recognitions

John W. Wilson Memorial Award, Research Council for Diagnostic and Prescriptive Mathematics, March 1986
Fulbright Distinguished Visitor, New Zealand Association for Research in Education, December 1987
Fulbright Senior Lecturer, Autonomous University of Barcelona, Spain, April 1989
Fulbright Senior Lecturer, University of the Andes, Colombia, March 1993
Fulbright Research Scholar, Gothenburg University, Sweden, August-November 1993
Regents Professorship, University of Georgia, 1993–present
Doctor Honoris Causa, Gothenburg University, Sweden, October 1995
Distinguished Visitor, Singapore Mathematics Society & Singapore Association of Mathematics Educators, February 2002
National Associate, National Academies of Science, 2002–present
Lifetime Achievement Award for Distinguished Service to Mathematics Education, National Council of Teachers of Mathematics, April 2003
2007 Felix Klein Medal honoring lifetime achievement in mathematics education awarded by the International Commission on Mathematical Instruction, Monterrey, Mexico, July 2008
Inaugural Fellow, American Educational Research Association, 2009–present
Member, National Academy of Education, 2010–present

Examples of Professional Service

American Educational Research Association

Special Interest Group for Research in Mathematics Education (Co-chair, 1972–1974)
Grants Program Governing Board, 2002–present

American Statistical Association

Working Group on Statistics in Mathematics Education Research, 2004–2007

College Board

Scholastic Aptitude Test Committee, 1977–1982 (Chair, 1980–1982)
Mathematical Sciences Advisory Committee, 1978–1989 (Chair, 1986–1989)
Council on Academic Affairs, 1986–1989 (Vice Chair, 1987–1989)
MAA/College Board Committee on Mutual Concerns, 1985–1989 (Co-chair, 1987–1989)
Equity 2000 National Mathematics Technical Assistance Committee, 1991–1997

International Commission on Mathematical Instruction (ICMI)

Executive Committee, 1987–1998 (Vice President, 1991–1998)
International Program Committee (IPC), 4th ICME, 1978–1980
IPC, 7th ICME, 1987–1992
IPC (Co-chair), ICMI Study on Research in Mathematics Education and Its Results, 1992–1996
IPC, 8th ICME, 1992–1996
IPC, Symposium for the Centennial of the ICMI, 2005–2008

Mathematical Association of America

Board of Governors, 1985–1987; 2006–2009
Common Ground Committee, 2004–2006

National Council of Teachers of Mathematics

Publications Committee, 1972–1976 (Chair, 1974–1976)
Task Force for the Research Agenda Project, 1980–1981
Advisory Board, Research Agenda Project, 1985–1989
Working Group for Assessment Standards for School Mathematics (Chair), 1993–1995
Commission on the Future of the Standards, 1996–2000

National Academy of Education

Focus Group on Learning and Instruction in School Subject Matter, 1989
Standing Review Committee, 2011–2014

National Assessment of Educational Progress

Advisory Panel, Mathematics Assessment, 1975–1980
Consultant on mathematics exercises, 1972, 1974, 1975–1982
Mathematics Item Development Panel, 1988
Mathematics Assessment Framework Planning Committee, 2000–2004

National Research Council

Mathematical Sciences Education Board, 1985–1986
Chair, New Approaches to Assessment Study Committee, 1990–1993
United States Commission on Mathematical Instruction, Ex-officio member, 1992–1998
Chair, Mathematics Learning Study, 1998–2001
Board on International Comparative Studies in Education, 2000–2003

National Science Foundation

Chair, Peer Oversight Committee for Research in Teaching and Learning, 1985–1986
Advisory Panel on NSF Centers of Excellence in Teacher Preparation, 1991
Expert Panel to Review the Mathematics Portfolio of NSF's EHR Directorate, 2003–2004

Office of Education Research and Innovation

OERI/RAND Mathematics Study Group, 2000–2003

Organisation for Economic Cooperation and Development (OECD)

U.S. Representative, Mathematics Expert Group for PISA, 2001–present

Selected Editorial Service

Chief Editor

Editor, *Journal for Research in Mathematics Education (JRME)*, 1982–1988
Editor, *JRME* Research Commentary section, 2004–2008

Editorial Board Member

Advisory Board, *Educatio Mathematica,* 1985–2004
Advisory Board, *For the Learning of Mathematics,* 1980–present
Advisory Board, *Uno: Revista de Didáctica de las Matemática,* 1994–present
Advisory Committee, *Zentralblatt für Didaktik der Mathematik,* 1970–present
Advisory Council, *Quadrante: Revista Teórica e de Investigação,* 1994–present
Comité de Redaction, *Recherches en Didactique des Mathématiques,* 1994–present
Comité editorial, *PNA: Revista de Investigatión en Didáctica de la Matemática,* 2007–present
Consejo Asesor, *Revista EMA: Investigación e Innovación en Educación Mathemática,* 1997–present
Contributing Editor, *Journal of Mathematical Behavior,* 1974–present
Editorial Board, *American Educational Research Journal,* Section on Social & Institutional Analysis, 2002–2005
Editorial Board, *Cognition and Instruction,* 1981–2003
Editorial Board, *Curricular and Policy Issues in Mathematics Education,* 1992–present
Editorial Board, *Educational Studies in Mathematics,* 1979–1989
Editorial Board, *Handbook of Research on Mathematics Teaching and Learning,* 1986–1992
Editorial Board, *International Journal for the History of Mathematics Teaching,* 2005–present
Editorial Board, *International Journal of Science and Mathematics Education,* 2002–2005
Editorial Committee (Chair), *Soviet Studies in Mathematics Education,* 1988–1992
International Editorial Board, Advances in Mathematics Education, 2009–present

Selected Publications

Sole Author

Kilpatrick, J. (1964). Cognitive theory and the SMSG program. *Journal of Research in Science Teaching, 2,* 247–251.
Kilpatrick, J. (1969). Problem-solving and creative behavior in mathematics. In J. W. Wilson & L. R. Carry (Eds.), *Reviews of recent research in mathematics education* (Studies in Mathematics Education, Vol. 19, pp. 153–187). Stanford: School Mathematics Study Group.
Kilpatrick, J. (1969). Problem solving in mathematics. *Review of Educational Research, 39,* 523–534.

Kilpatrick, J. (1971). Some implications of the International Study of Achievement in Mathematics for mathematics educators. *Journal for Research in Mathematics Education, 2*, 164–171.

Kilpatrick, J. (1971). Individual differences that might influence the effectiveness of instruction in mathematics. In H. Bauersfeld, M. Otte, & H.-G. Steiner (Eds.), *Schriftenreihe des IDM* (Vol. 4, pp. 67–82). Bielefeld: Universität Bielefeld, Institut für Didaktik der Mathematik.

Kilpatrick, J. (1978). Research on problem solving in mathematics. *School Science and Mathematics, 78*, 189–192.

Kilpatrick, J. (1978). Theories of learning and their implications for teaching informatics and mathematics. In D. C. Johnson & J. D. Tinsley (Eds.), *Informatics and mathematics in secondary schools: Impacts and relationships* (pp. 93–99). Amsterdam: North-Holland.

Kilpatrick, J. (1981). Mathematics examinations at the end of secondary school in the USA. *Zentralblatt für Didaktik der Mathematik, 13*, 181–185.

Kilpatrick, J. (1981). The reasonable ineffectiveness of research in mathematics education. *For the Learning of Mathematics, 2*(2), 22–29.

Kilpatrick, J. (1981). Research on mathematical learning and thinking in the United States. *Recherche en Didactique des Mathématiques, 2*, 363–379.

Kilpatrick, J. (1985). Reflection and recursion. *Educational Studies in Mathematics, 16*, 1–26. (Also published in Carss, M. (Ed.), *Proceedings of the Fifth International Congress on Mathematical Education* (pp. 7–29). Boston: Birkhäuser.)

Kilpatrick, J. (1985). A retrospective account of the past twenty-five years of research on teaching mathematical problem solving. In E. A. Silver (Ed.), *Teaching and learning mathematical problem solving: Multiple research perspectives* (pp. 1–15). Hillsdale: Erlbaum.

Kilpatrick, J. (1987). George Polya's influence on mathematics education. *Mathematics Magazine, 60*, 299–300.

Kilpatrick, J. (1987). Is teaching teachable? George Polya's views on the training of mathematics teachers. In F. R. Curcio (Ed.), *Teaching and learning: A problem-solving focus* (pp. 85–97). Reston: National Council of Teachers of Mathematics.

Kilpatrick, J. (1987). Problem formulating: Where do good problems come from? In A. H. Schoenfeld (Ed.), *Cognitive science and mathematics education* (pp. 123–147). Hillsdale: Erlbaum.

Kilpatrick, J. (1987). What constructivism might be in mathematics education. In J. C. Bergeron, N. Herscovics, & C. Kieran (Eds.), *Proceedings of the 11th International Conference for the Psychology of Mathematics Education* (Vol. 1, pp. 3–37). Montreal: University of Montreal.

Kilpatrick, J. (1988). Change and stability in research in mathematics education. *Zentralblatt für Didaktik der Mathematik, 20*, 202–204.

Kilpatrick, J. (1988). Educational research: Scientific or political? *Australian Educational Researcher, 15*(2), 13–28.

Kilpatrick, J. (1991). Mathematics education: Introduction. In A. Lewy (Ed.), *International Encyclopedia of Curriculum* (pp. 819–820). Oxford: Pergamon.

Kilpatrick, J. (1991). Problem-solving in mathematics. In A. Lewy (Ed.), *International Encyclopedia of Curriculum* (pp. 847–848). Oxford: Pergamon.

Kilpatrick, J. (1992). "America is likewise bestirring herself": A century of mathematics education as viewed from the United States. In I. Wirszup and R. Streit (Eds.), *Developments in school mathematics education around the world* (Vol. 3, pp. 133–145). Reston: National Council of Teachers of Mathematics.

Kilpatrick, J. (1992). A history of research in mathematics education. In D. Grouws (Ed.), *Handbook of research on mathematics teaching and learning* (pp. 3–38). New York: Macmillan. (Reprinted, in Spanish, in *Educación matemática e investigación*, by J. Kilpatrick, L. Rico, & M. Sierra (1994, pp. 13–96). Madrid: Editorial Síntesis).

Kilpatrick, J. (1992). Scattering, storing, shaping: Journals in mathematics education. *Nämnaren, 18*(3/4), 16–23.

Kilpatrick, J. (1992). Some issues in the assessment of mathematical problem solving. In J. Ponte (Ed.), *Mathematical problem solving and new information technologies: Research in contexts of practice* (pp. 37–44). Berlin: Springer.

Kilpatrick, J. (1993). Beyond face value: Assessing research in mathematics education. In G. Nissen & M. Blomhøj (Eds.), *Criteria for scientific quality and relevance in the didactics of mathematics* (pp. 15–34). Roskilde: IMFUFA, Roskilde University.

Kilpatrick, J. (1993). The chain and the arrow: From the history of mathematics assessment. In M. Niss (Ed.), *Investigations into assessment in mathematics education: An ICMI study* (pp. 31–46). Dordrecht: Kluwer.

Kilpatrick, J. (1994). History of mathematics education. In T. Husén & T. N. Postlethwaite (Eds.), *International encyclopedia of education* (2nd. ed., pp. 3643–3647). Oxford: Pergamon.

Kilpatrick, J. (1994). Mathematics instruction: Contemporary research. In T. Husén & T. N. Postlethwaite (Eds.), *International encyclopedia of education* (2nd. ed., pp. 3647–3652). Oxford: Pergamon.

Kilpatrick, J. (1994). Vingt ans de didactique française depuis les USA (Twenty years of French didactics seen from the USA). In M. Artique, R. Gras, C. Laborde, & P. Tavignot (Eds.), *Vingt ans de didactique des mathématiques en France* (pp. 84–96). Grenoble: La Pensée Sauvage.

Kilpatrick, J. (1995). Curriculum change locally and globally. In R. P. Hunting, G. E. Fitzsimons, P. C. Clarkson, & A. J. Bishop (Eds.), *Regional collaboration in mathematics education 1995* (pp. 19–29). Melbourne: Monash University, Faculty of Education.

Kilpatrick, J. (1966). Fincando estacas: Uma tentativa de demarcar a educação matemática como campo profissional e científico (Staking claims: An attempt to delimit mathematics as a professional and scientific field). *Zetetiké, 4*(5), 99–120.

Kilpatrick, J. (1997). Confronting reform. *American Mathematical Monthly, 104,* 955–962.

Kilpatrick, J. (1998). The research culture of U.S. mathematics education. In T. Lingefjärd & G. Dahland (Eds.), *Research in mathematics education: A report from a follow-up conference after PME 1997* (pp. 39–53). Gothenburg: University of Gothenburg, Department of Subject Matter Didactics, Section of Mathematics Education.

Kilpatrick, J. (1999). Ich bin Europäisch (I am a European). In I. Schwank (Ed.), *European research in mathematics education* (Vol. 1, pp. 51–70). Osnabrück: Forshungsinstitut für Mathematikdidaktik.

Kilpatrick, J. (2000). Research in mathematics education across two centuries. In M. A. Clements, H. H. Tairab, & W. K. Yoong (Eds.), *Science, mathematics and technical education in the 20th and 21st centuries* (pp. 79–93). Gadong: Universiti Brunei Darussalam.

Kilpatrick, J. (2001). Recalling David Wheeler. *For the Learning of Mathematics, 21*(2), 16–17.

Kilpatrick, J. (2001). Where's the evidence? *Journal for Research in Mathematics Education, 32*, 421–427.

Kilpatrick, J. (2001). Understanding mathematical literacy: The contribution of research. *Educational Studies in Mathematics, 47*, 101–116.

Kilpatrick, J. (2003). Twenty years of French didactique viewed from the United States. *For the Learning of Mathematics, 23*(2), 23–27.

Kilpatrick, J. (2003). Scientific solidarity today and tomorrow. In D. Coray, F. Furinghetti, H. Gispert, B. Hodgson, & G. Schubring (Eds.), *One hundred years of L'Enseignement Mathématique: Moments of mathematics education in the twentieth century* (Monographie No. 39 de L'Enseignement Mathématique, pp. 317–330).

Kilpatrick, J. (2003). What works? In S. L. Senk & D. R. Thompson (Eds.), *Standards-based school mathematics curricula: What are they? What do students learn?* (pp. 471–488). Mahwah: Erlbaum.

Kilpatrick, J. (2004). Centers for learning and teaching. In R. Strässer, G. Brandell, B. Grevholm, & O. Helenius (Eds.), *Educating for the future: Proceedings of an International Symposium on Mathematics Teacher Education* (pp. 143–157). Stockholm: Royal Swedish Academy of Sciences.

Kilpatrick, J. (2004). Synthesizing research on learning mathematics. In A. Tengstrand (Ed.), *Proceedings of the Nordic Pre-conference to ICME 10 at Växjö University, Sweden, May 9–11, 2003* (pp. 1–11). Växjö: Växjö University.

Kilpatrick, J. (2004). Variables and methodologies in research on problem solving. In T. P. Carpenter, J. A. Dossey, & J. L. Koehler (Eds.), *Classics in mathematics education research* (pp. 41–47). Reston: National Council of Teachers of Mathematics. (Reprinted from *Mathematical problem solving: Papers from a research workshop*, pp. 7–20, by L. L. Hatfield & D. A. Bradbard, Eds., 1978, Columbus: ERIC Clearinghouse for Science, Mathematics, and Environmental Education).

Kilpatrick, J. (2005). A critique of impure unreason. In L. Santos, A. P. Canavarro, & J. Brocado (Eds.), *Mathematics education: Paths and Crossroads. International meeting in honour of Paulo Abrantes, 14–15 July, Lisbon* (pp. 109–119). Lisbon: Etigrafe.

Kilpatrick, J. (2006). A golden means to teaching mathematics effectively (Review of *The middle path in math instruction: Solutions for improving math education*). *Journal for Research in Mathematics Education, 37,* 256–259.

Kilpatrick, J. (2007). O papel e o alcance das revistas na educação matemática [The role and scope of journals in mathematics education]. *Educação e Matemática, 91,* 66–72.

Kilpatrick, J. (2007). Developing common sense in teaching mathematics. In U. Gellert & E. Jablonka (Eds.), *Mathematisation and demathematisation: Social, philosophical and educational ramifications* (pp. 161–169). Rotterdam: Sense Publishers.

Kilpatrick, J. (2008). The development of mathematics education as an academic field. In M. Menghini, F. Furinghetti, L. Giacardi, & F. Arzarello (Eds.), *The first century of the International Commission on Mathematical Instruction (1908–2008): Reflecting and shaping the world of mathematics education* (pp. 25–39). Rome: Istituto della Enciclopedia Italiana.

Kilpatrick, J. (2008). Research in mathematics education: Learning from each other. In R. Luengo González, B. Gómez Alfonso, M. Camacho Machín, & L. J. Blanco Nieto (Eds.), *Investigación en educación matemática XII* (pp. 79–92). Badajoz: Indugrafic Artes Gráficas.

Kilpatrick, J. (2008). Practicing research and researching practice. In P. Clarkson & N. Presmeg (Eds.), *Critical issues in mathematics education: Major contributions of Alan Bishop* (pp. 205–212). New York: Springer.

Kilpatrick, J. (2010). Influences of Soviet research in mathematics education. In A. Karp & B. R. Vogeli (Eds.), *Russian mathematics education: History and world significance* (Series on Mathematics Education, Vol. 4, pp. 359–368). Hackensack: World Scientific.

Kilpatrick, J. (2011). A look back … Pólya on mathematical abilities. *The Mathematics Educator, 21*(1), 3–8. Full interview transcript published in Portuguese in H. Guimarães, (2010), Jeremy Kilpatrick: entrevista a George Pólya (Jeremy Kilpatrick: Interview with George Pólya), *Quadrante, 19*(2), 103–119. Also published in H. Guimarães, (2011), Pólya e as capacidades matemáticas (Pólya and mathematical abilities), *Educação e Matemática, 114,* 28–36.

Kilpatrick, J. (2011). Research in the learning and teaching of mathematics: What can researchers in education and mathematics learn from each other? In J. Emanuelsson, L. Fainsilber, J. Häggström, A. Kullberg, B. Lindström, & M. Löwing (Eds.), *Voices on learning and instruction in mathematics* (pp. 63–77). Gothenburg: University of Gothenburg, National Center for Mathematics Education.

Kilpatrick, J. (2011, Spring-Summer). Slouching toward a national curriculum. *Journal of Mathematics Education at Teachers College, 2,* 8–17.

Kilpatrick, J. (2012). The new math as an international phenomenon. *ZDM: The International Journal on Mathematics Education, 44,* 563–571.

Kilpatrick, J. (2012). U.S. mathematicians and the new math movement. In K. Bjarnadóttir, F. Furinghetti, J. M. Matos, & G. Schubring (Eds.), *"Dig where you stand" 2: Proceedings of the Second International Conference on the History of Mathematics Education* (pp. 216–226). Lisbon: Unidade de Investigação Educação e Desenvolvimento, Faculdade de Ciências e Tecnologia da Universidade Nova de Lisboa.

Kilpatrick, J. (2013). Introduction to Section D: International perspectives on mathematics education. In M. A. (Ken) Clements, A. J. Bishop, C. Keitel, J. Kilpatrick, & F. K. S. Leung (Eds.), *Third international handbook of mathematics education* (pp. 791–795). New York: Springer.

Kilpatrick, J. (2013, Spring-Summer). Leading people: Leadership in mathematics education. *Journal of Mathematics Education at Teachers College, 2,* 7–14.

Kilpatrick, J. (2014). Competency frameworks in mathematics education. In S. Lerman (Ed.), *Encyclopedia of mathematics education*. Heidelberg: Springer.

Kilpatrick, J. (2014). History of research in mathematics education. In S. Lerman (Ed.), *Encyclopedia of mathematics education*. Heidelberg: Springer.

Kilpatrick, J. (2013). Mathematics education. In G. McCulloch & D. Crook (Eds.), *International encyclopedia of education*. London: Routledge.

Co-author

Atkin, J. M., Kilpatrick, J., Bianchini, J. A., Helms, J. V., & Holthuis, N. I. (1997). The changing conceptions of science, mathematics, and instruction. In S. A. Raizen & E. D. Britton (Eds.), *Bold ventures, Vol. 1: Patterns among innovations in science and mathematics education* (pp. 43–72). Dordrecht: Kluwer.

Ball, D. L, Ferrini-Mundy, J., Kilpatrick, J., Milgram, R. J., Schmid, W., & Schaar, R. (2005). Reaching for common ground in K–12 mathematics education. *Notices of the American Mathematical Society, 52,* 1055–1058. (Reprinted in Winter 2007 in *Leadership Information, 6*(1), 13–15).

Branca, N. A., & Kilpatrick, J. (1972). The consistency of strategies in the learning of mathematical structures. *Journal for Research in Mathematics Education, 3,* 132–140.

Clements, M. A. (Ken), Bishop, A. J., Keitel, C., Kilpatrick, J., & Leung, F. K. S. (2013). From the few to the many: Historical perspectives on who should learn mathematics. In M. A. (Ken) Clements, A. J. Bishop, C. Keitel, J. Kilpatrick, & F. K. S. Leung (Eds.), *Third international handbook of mathematics education* (pp. 7–40). New York: Springer.

Gieger, J. L., & Kilpatrick, J. (1999). Mathematically promising students: National trends and international comparisons. In L. J. Sheffield (Ed.), *Developing mathematically promising students* (pp. 27–41). Reston: National Council of Teachers of Mathematics.

Hancock, L., & Kilpatrick, J. (1993). Effects of mandated testing on instruction. Commissioned paper in Mathematical Sciences Education Board, National Research Council, *Measuring what counts: A conceptual guide to mathematics assessment* (pp. 149–174). Washington, DC: National Academy Press.

Herbst, P., & Kilpatrick, J. (1999). *Pour lire* Brousseau [Reading Brousseau]. *For the Learning of Mathematics, 19*(1), 3–10.

Hiebert, J., Kilpatrick, J., & Lindquist, M. M. (2001). Improving U.S. doctoral programs in mathematics education. In R. E. Reys & J. Kilpatrick (Eds.), *One field, many paths: U.S. doctoral programs in mathematics education* (pp. 153–159). Washington, DC: Conference Board of the Mathematical Sciences.

Howson, G., Keitel, C., & Kilpatrick, J. (1981). *Curriculum development in mathematics*. Cambridge: Cambridge University Press. [Japanese edition, 1987]

Kang, W., & Kilpatrick, J. (1992). Didactic transposition in mathematics textbooks. *For the Learning of Mathematics, 12*(1), 2–7.

Keitel, C., & Kilpatrick, J. (1998). The rationality and irrationality of international comparative studies. In G. Kaiser, E. Luna, & I. Huntley (Eds.), *International comparisons in mathematics education* (pp. 241–256). London: Falmer. (Reprinted in A. J. Bishop (Ed.), (2010). *Mathematics education* (Vol. 4, pp. 166–180). London: Routledge).

Keitel, C., & Kilpatrick, J. (2005). Mathematics education and common sense. In J. Kilpatrick, C. Hoyles, & O. Skovsmose (Eds.), Meaning in mathematics education (pp. 105–128). Dordrecht: Kluwer.

Kilpatrick, J., & Davis, R. B. (1993). Computers and curriculum change in mathematics. In C. Keitel & K. Ruthven (Eds.), *Learning from computers: Mathematics education and technology* (pp. 203–221). Berlin: Springer.

Kilpatrick, J., & Gieger, J. L. (2000). The performance of students taking advanced mathematics courses. In P. A. Kenney & E. A. Silver (Eds.), *Results from the seventh mathematics assessment of the National Assessment of Educational Progress* (pp. 377–409). Reston: National Council of Teachers of Mathematics.

Kilpatrick, J., Hancock, L., Mewborn, D. L., & Stallings, L. (1996). Teaching and learning cross-country mathematics: A story of innovation in precalculus. In S. A. Raizen & E. D. Britton (Eds.), *Bold ventures, Vol. 3: Case studies of U.S. innovations in mathematics education* (pp. 133–243). Dordrecht: Kluwer.

Kilpatrick, J., & Izsák, A. (2008). A history of algebra in the school curriculum. In C. E. Greenes & R. Rubenstein (Eds.), *Algebra and algebraic thinking in school mathematics* (2008 Yearbook of the National Council of Teachers of Mathematics, pp. 3–18). Reston: NCTM.

Kilpatrick, J., & Johansson, B. (1993). Standardized mathematics testing in Sweden: The legacy of Frits Wigforss. *Nordic Studies in Mathematics Education, 2*, 6–30.

Kilpatrick, J., & Lingefjärd, T. (1996). Vildmarksmatematik [Cross-country mathematics]. *Nämnaren, 23*(2), 17–25.

Kilpatrick, J., Mesa, V., & Sloane, F. (2007). U.S. algebra performance in an international context. In T. Loveless (Ed.), *Lessons learned: What international assessments tell us about math achievement* (pp. 85–126). Washington, DC: Brookings Institution Press.

Kilpatrick, J., & Moura, E. B. (1999). Reflexões sobre os Standards [Reflections on the Standards]. *Educação e Matemática, 55*, 43–46.

Kilpatrick, J., & Radatz, H. (1983). How teachers might make use of research on problem solving. *Zentralblatt für Didaktik der Mathematik, 15*, 151–155.

Kilpatrick, J., & Sierpinska, A. (2001). A report on the ICMI Study: "What is research in mathematics education, and what are its results?" *L'Enseignement Mathématique, 47,* 409–411.

Kilpatrick, J., & Silver, E. A. (2000). Unfinished business: Challenges for mathematics educators in the next decades. In M. J. Burke & F. R. Curcio (Eds.), *Learning mathematics for a new century* (2000 Yearbook of the National Council of Teachers of Mathematics, pp. 223–235). Reston: National Council of Teachers of Mathematics.

Kilpatrick, J., & Stanic, G. M. A. (1995). Paths to the present. In I. M. Carl (Ed.), *Seventy-five years of progress: Prospects for school mathematics* (pp. 3–17). Reston: National Council of Teachers of Mathematics.

Lingefjärd, T., & Kilpatrick, J. (1998). Authority and responsibility when learning mathematics in a technology-enhanced environment. In D. Tinsley & D. C. Johnson (Eds.), *Information and communications technologies in mathematics* (pp. 233–236). London: Chapman & Hall.

Mewborn, D. S., & Kilpatrick, J. (2010). Removing disparities in mathematics achievement. In R. I. Charles & F. K. Lester, Jr. (Eds.), *Teaching and learning mathematics: Translating research for school administrators* (pp. 7–12). Reston: National Council of Teachers of Mathematics.

Mewborn, D. S., & Kilpatrick, J. (2010). School mathematics for the twenty-first century. In R. I. Charles & F. K. Lester, Jr. (Eds.), *Teaching and learning mathematics: Translating research for school administrators* (pp. 1–6). Reston: National Council of Teachers of Mathematics.

Nipper, K., Ricks, T., Kilpatrick, J., Mayhew, L., Thomas, S., Kwon, N. Y., Klerlein, J. T., & Hembree, D. (2011). Teacher tensions: Expectations in a professional development institute. *Journal of Mathematics Teacher Education, 14,* 375–392.

Polya, G., & Kilpatrick, J. (2009). *The Stanford mathematics problem book: With hints and solutions* (Dover Books on Mathematics). Mineola: Dover. (Original work published 1974)

Schneider, B., Carnoy, M., Kilpatrick, J., Schmidt, W., & Shavelson, R. (2007). *Estimating causal effects using experimental and observational designs.* Washington, DC: American Educational Research Association.

Schoenfeld, A., & Kilpatrick, J. (2008). Toward a theory of proficiency in teaching mathematics. In D. Tirosh & T. Wood (Eds.), *Tools and processes in mathematics teacher education* (pp. 321–354). Rotterdam: Sense Publishers.

Sierpinska, A., & Kilpatrick, J. (1998). Continuing the search. In A. Sierpinska & J. Kilpatrick (Eds.), *Mathematics education as a research domain: A search for identity* (New ICMI Study Series, Vol. 4, Book 2, pp. 527–548). Dordrecht: Kluwer.

Silver, E. A., & Kilpatrick, J. (1994). E pluribus unum: Challenges of diversity in the future of mathematics education research. *Journal for Research in Mathematics Education, 25,* 734–754.

Silver, E. A., & Kilpatrick, J. (1988). Testing mathematical problem solving. In R. I. Charles & E. A. Silver (Eds.), *Research agenda for mathematics education: Vol. 3.*

The teaching and assessing of mathematical problem solving (pp. 178–186). Hillsdale: Erlbaum.

Silver, E. A., Kilpatrick, J., & Schlesinger, B. (1990). *Thinking through mathematics: Fostering inquiry and communication in mathematics classrooms.* New York: College Entrance Examination Board.

Stanic, G. M. A., & Kilpatrick, J. (1988). Historical perspectives on problem solving in the mathematics curriculum. In R. I. Charles & E. A. Silver (Eds.), *Research agenda for mathematics education: Vol. 3. The teaching and assessing of mathematical problem solving* (pp. 1–22). Hillsdale: Erlbaum.

Stanic, G. M. A., & Kilpatrick, J. (1992). Mathematics curriculum reform in the United States: A historical perspective. *International Journal of Educational Research, 17,* 407–437.

Wilson, P. S., & Kilpatrick, J. (2005, Fall). Mathematics education as a commitment. *Newsletter of the Mathematicians and Education Reform Forum, 18*(1), 4–5.

Wilson, J. W., & Kilpatrick, J. (1989). Theoretical issues in the development of calculator-based mathematics tests. In J. W. Kenelly (Ed.), *The use of calculators in the standardized testing of mathematics* (MAA Notes No. 12, pp. 7–15). New York: College Entrance Examination Board.

Editor or Co-editor

Editor (with Izaak Wirszup). (1969–1972). *Soviet studies in the psychology of learning and teaching mathematics* (Vols. 1–6). Stanford: School Mathematics Study Group.

Editor (with Izaak Wirszup, Edward G. Begle, & James W. Wilson). (1972–1975). *Soviet studies in the psychology of learning and teaching mathematics* (Vols. 7–14). Stanford: School Mathematics Study Group.

Editor (with Izaak Wirszup). Krutetskii, V. A. (1976). *The psychology of mathematical abilities in schoolchildren.* Chicago: University of Chicago Press.

Editor (with James W. Wilson). Begle, E. G. (1979). *Critical variables in mathematics education: Findings from a survey of the empirical literature.* Washington, DC: Mathematical Association of America & National Council of Teachers of Mathematics.

Editor (with Marilyn Zweng, Thomas Green, Henry Pollak, & Marilyn Suydam). (1983). *Proceedings of the Fourth International Congress on Mathematical Education.* Boston: Birkhäuser.

Editor (with Alan Bell & Brian Low). (1985). *Theory, research and practice in mathematical education.* Nottingham: University of Nottingham, Shell Centre for Mathematical Education.

Editor (with Pearla Nesher). (1990). *Mathematics and cognition: A research synthesis by the International Group for the Psychology of Mathematics Education.* Cambridge: Cambridge University Press.

Editor. Davydov, V. V. (1990). *Types of generalization in instruction: Logical and psychological problems in the structuring of school curricula* (Soviet Studies in Mathematics Education, Vol. 2). Reston: National Council of Teachers of Mathematics.

Editor. Dubrovina, I. V., & Shapiro, S. I. (1992). *Problems in the psychology of abilities: A collection of articles* (Soviet Studies in Mathematics Education, Vol. 8). Chicago: University of Chicago School Mathematics Project.

Editor (with Alan Bishop, Ken Clements, Christine Keitel, & Colette Laborde). (1996). *International handbook of mathematics education.* Dordrecht: Kluwer. [36 chapters in all; J. Kilpatrick edited 13 chapters in section on curriculum.]

Editor (with Alan Bishop, Ken Clements, Christine Keitel, & Frederick Leung). (2003). *International handbook of mathematics education* (2nd ed.). Dordrecht: Kluwer. (24 chapters in all; J. Kilpatrick edited 7 chapters in section on research.)

Editor (with Anna Sierpinska). (1998). *Mathematics education as a research domain: A search for identity* (New ICMI Study Series, Vol. 4, Books 1 & 2). Dordrecht: Kluwer.

Editor (with Robert E. Reys). (2001). *One field, many paths: U.S. doctoral programs in mathematics education.* Washington, DC: Conference Board of the Mathematical Sciences.

Editor (with Jane Swafford & Brad Findell). (2001). *Adding it up: Helping children learn mathematics.* Washington, DC: National Academy Press.

Editor (with Jane Swafford). (2002). *Helping children learn mathematics.* Washington, DC: National Academy Press.

Editor (with George M. A. Stanic). (2003). *A history of school mathematics.* Reston: National Council of Teachers of Mathematics.

Editor (with W. Gary Martin & Deborah E. Schifter). (2003). *A research companion to principles and standards for school mathematics.* Reston: National Council of Teachers of Mathematics.

Editor (with Celia Hoyles & Ole Skovsmose). (2005). *Meaning in mathematics education.* Dordrecht: Kluwer.

Editor (with M. A. (Ken) Clements, Alan J. Bishop, Christine Keitel, & Frederick K. S. Leung). (2013). *Third international handbook of mathematics education.* New York: Springer.

Principal author and editor. Mathematical Sciences Education Board, National Research Council. (1993). *Measuring what counts: A conceptual guide for mathematics assessment.* Washington, DC: National Academy Press.

Principal author, editor, and chair of Working Group. National Council of Teachers of Mathematics (1995). *Assessment standards for school mathematics.* Reston: NCTM.

Biographical Sketches of Contributors

Emeritus Professor Alan Bishop was Professor of Education and Associate Dean at Monash University between 1992 and 2002 after spending the earlier part of his life in the UK. In 1969, after completing his PhD, he was appointed Lecturer in the Department of Education at Cambridge University, UK, where he worked for 23 years. He edited (from 1978 to 1990) the international research journal *Educational Studies in Mathematics* published by Kluwer (now Springer), and he has been an Advisory Editor since 1990. He is Managing Editor of the research book series *Mathematics Education Library* also published by Kluwer/Springer (1980 – present). He was the Chief Editor of two *International Handbooks of Mathematics Education* (1996 and 2002) published by Kluwer/Springer and joint Editor of the *Third International Handbook of Mathematics Education* (2012) also published by Springer. He was the sole Editor of the *Handbook on Mathematics Education* published by Routledge (2003).

João Pedro da Ponte is Professor and Director of the Institute of Education of the University of Lisbon. He has done his doctoral studies at the University of Georgia (USA) and has been a visiting Professor at several universities including San Diego (USA), UNICAMP (Brazil), and Granada (Spain). His current main research interests are mathematics teaching practices and teacher education, with a special focus on the teaching of algebra, rational numbers, and statistics and in the students' development of mathematics reasoning. He coordinated a government report about preservice teacher education (2006) and a new mathematics curriculum for basic education (2007) and collaborates regularly with the Portuguese association of teachers of mathematics. He has supervised more than 20 PhD dissertations and is author and coauthor of several books and articles in national and international journals and associate editor of the *Journal of Mathematics Teacher Education*.

Pat Herbst received his doctorate from the University of Georgia in 1998 working under Jeremy Kilpatrick's direction. Since 1999, he has been on the faculty at the

© Springer International Publishing Switzerland 2015
E. Silver, C. Keitel-Kreidt (eds.), *Pursuing Excellence in Mathematics Education*, Mathematics Education Library, DOI 10.1007/978-3-319-11952-6

University of Michigan School of Education, where he is now Professor of Education and Mathematics. Since 2001, Herbst has directed an active, well-funded research lab, the GRIP (Geometry, Reasoning, and Instructional Practices), which has served as context for the development of several doctoral students and postdoctoral fellows. In 2011, Herbst and his associates opened the LessonSketch site (www.lessonsketch.org), which includes facilities for researchers to create multimedia questionnaires and distribute them online. With those facilities, the group has been gathering evidence to support Herbst & Chazan's theory of practical rationality as well as enabling other researchers to do research and development online.

Thomas Lingefjärd received his Ph.D. from the University of Georgia in 2000, working under Jeremy Kilpatrick's direction. Since 1988, he has been on the School of Education within the University of Gothenburg, where he was once a student, was employed as a lecturer, became senior lecturer and associate professor, and is now Professor of Mathematics Education. Lingefjärd is a textbook author, his research is mainly focused on higher education, and he enjoys rebuilding summer houses as well as geocaching and downhill skiing as outdoor activities.

John Mason is emeritus professor of mathematics and mathematics education at the Open University in England. After a BSc at Trinity College, Toronto, in Mathematics and an MSc at Massey College, Toronto, he went to Madison, Wisconsin, where he encountered Polya's film "Let Us Teach Guessing" and completed a PhD in Combinatorial Geometry. The film released a style of teaching he had experienced at high school from his mathematics teacher Geoff Steel, and his teaching changed overnight. Now, retired from the Open University after 40 years of preparing distance learning material for teachers of mathematics at all phases, he continues to work on mathematical problems, most of which arise from pedagogical concerns and issues, and on the nature and role of attention in teaching and learning mathematics. Among his many publications are *Thinking Mathematically* and *Researching Your Own Practice: The Discipline of Noticing*.

Vilma Mesa is Assistant Professor of Education at the University of Michigan. She investigates the role that resources play in developing teaching expertise in undergraduate mathematics, specifically at community colleges and in inquiry-based learning classrooms. She has conducted several analyses of textbooks and evaluation projects on the impact of innovative mathematics teaching practices for students in science, technology, engineering, and mathematics. She has a B.S. in computer sciences and a B.S. in mathematics from the University of Los Andes in Bogotá, Colombia, and a master's and a Ph.D. in mathematics education from the University of Georgia.

Pearla Nesher is a Professor of mathematics education at the The University of Haifa and former Dean of Haifa School of Education. She got her education at the Hebrew University, Jerusalem, and Harvard University, Cambridge, Massachusetts. Prof. Nesher has devoted more than 25 years to research in mathematics education from a cognitive perspective. She is expert in the area of problem solving and has published numerous papers on this topic. Currently, she is studying brain activity while engaging in arithmetical actions.

She served 4 years as the Chief Scientist of the Ministry of Education in Israel. In that position, she was responsible for enhancing learning and school achievement, assisting to open higher education for more students in Israel. She served as the president of the international Group for the Psychology of Learning Mathematics (PME) and was a visiting scholar at MIT, Harvard, and Pittsburgh University.

Edward A. Silver (easilver@umich.edu) is the William A. Brownell Collegiate Professor of Education and Professor of Mathematics at the University of Michigan. His main teaching and advising responsibilities involve doctoral students in mathematics education. His scholarly interests include the study of mathematical thinking, especially mathematical problem solving and problem posing; the design and analysis of intellectually engaging and equitable mathematics instruction for students; innovative methods of assessing and reporting mathematics achievement; and effective models for enhancing the knowledge of teachers of mathematics. He has published broadly on these topics and has directed or codirected a number of projects in mathematics education related to these themes. From 2000 to 2004, he served as editor of the *Journal for Research in Mathematics Education*, and he served from 2008 to 2010 as coeditor of *The Elementary School Journal*. As a graduate student at Teachers College, Columbia University, in the 1970s, he had the good fortune to study under Jeremy Kilpatrick.

James W. Wilson is professor of mathematics education at the University of Georgia, where he served 23 years as the head of the Mathematics Education Department and the graduate coordinator. There were two additional 3 year terms as graduate coordinator. He has mentored more than 55 doctoral students. He joined the faculty at the University of Georgia in 1968 after receiving his doctorate at Stanford University, where he was a classmate of Jeremy Kilpatrick. He also holds master's degrees in mathematics from Kansas State Teachers College, Notre Dame, and Stanford. Wilson's scholarship has focused on problem solving, assessment, and technology; he has directed a number of national, regional, and state initiatives aimed at improving mathematics teaching consistent with these themes. From 1976 to 1982, he served as editor of the *Journal for Research in Mathematics Education*. He also served as an elected member of the Board of Directors of the National Council for Teachers of Mathematics. In 2001, Wilson received the Lifetime Achievement Award from the National Council for Teachers of Mathematics, and in 2008, he was recipient of the Gladys M. Thomason Award for Distinguished Service from the Georgia Council of Teachers of Mathematics.

Index

© Springer International Publishing Switzerland 2015
E. Silver, C. Keitel-Kreidt (eds.), *Pursuing Excellence in Mathematics
Education*, Mathematics Education Library, DOI 10.1007/978-3-319-11952-6